OUR SCIENTIFIC
HERITAGE

For Sylvia,
my fellow rambler

OUR SCIENTIFIC
HERITAGE

AN A–Z OF
GREAT BRITAIN
AND IRELAND

TREVOR I. WILLIAMS

SUTTON PUBLISHING

First published in 1996 by
Sutton Publishing Limited · Phoenix Mill
Thrupp · Stroud · Gloucestershire · GL5 2BU

British Library Cataloguing in Publication Data
A catalogue record for this book is available from the British Library

ISBN 0-7509-0820-3

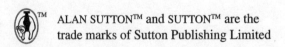

Typeset in 10/12 Sabon.
Typesetting and origination by
Sutton Publishing Limited.
Printed in Great Britain by
Butler & Tanner, Frome, Somerset.

CONTENTS

PREFACE

Wherever one may be in Britain there are countless reminders of all aspects of our rich cultural heritage. This book has been written to direct attention to one aspect in particular – achievements in science, technology and medicine. It is neither a reference book nor a tourist guide but rather one in which the reader may browse at home or consult when travelling to gain additional background to local sightseeing. Some of the places listed represent substantial survivals, such as Brunel's famous suspension bridge over the Avon Gorge at Bristol, or Newton's home at Woolsthorpe. Others may be memorials commemorating great achievements, such as the obelisk on the Lizard Peninsula marking the site of Marconi's historic transatlantic wireless transmission in 1901. Others again may be grand memorials in Westminster Abbey or simple ones in country churchyards such as Edward Jenner's grave at St Mary's, Berkeley. Some allusions, such as those to the Lunar Society which flourished in Birmingham in the 18th century, are necessarily general, for although highly influential in their day they have left little or nothing in the way of tangible relics.

The entries range over science, technology, and medicine in the widest sense and as archaeology is now firmly established as a science in its own right, it has been thought appropriate to include it within the remit. A fair number of the most outstanding and accessible archaeological sites have therefore been included, mainly stone circles, graves, hillforts and a few Roman settlements. The list could have been extended almost indefinitely, for such survivals are numbered in hundreds, but to have added many more would have unbalanced the book.

Over the centuries, power – wind, water and steam – has been fundamental to industrial development, and all three are strongly represented in these pages. Again, however, the treatment has necessarily been selective. The post-war years have seen a growing national enthusiasm for the restoration and preservation of windmills, watermills, and stationary steam engines and locomotives. Many narrow-gauge industrial railways have also been revived, as well as some sections of the old mainline system closed in the Beeching cuts. In all, they are too numerous to list in full: many are purely tourist attractions, and the emphasis here is on those which have associated museums and exhibitions.

Museums in general naturally figure prominently but the space accorded to each is not to be taken as a measure of stature. The great comprehensive national collections – such as the Science Museum in London and the Birmingham Museum of Science and Industry – have largely been left to speak for themselves. Their fame has spread far and the visitor will find a wealth of literature available. Rather, attention has been directed to the many excellent specialist collections – for example, the National Waterways Museum in Gloucester and the Llanwernog Silver-Lead Mine Museum at Ponterwyd –

which are relatively little known. Additionally, mention is made of the many municipal and other museums of a general nature which include substantial sections devoted wholly to some aspect of science and technology, often with local connotations.

On this basis, a surprising number of independent museums have been identified, many of them created since the 1950s. In many the standard is high and they attract sufficient visitors to be self-sustaining. Some, however, have been less successful and have fallen by the wayside. A cautionary note must therefore be sounded: the listing of a museum, especially a small one in a remote area, is not an absolute guarantee that it still exists. With this reservation, all those listed are to some extent accessible to the public but as opening times vary from season to season and year to year no attempt has been made to indicate these nor what charge if any, may be made.

In the nature of things, the geographical pattern of entries is uneven. The ancient centres of learning – like Oxford, Cambridge, London, Edinburgh and Glasgow – yield a rich harvest. So, too, in a different way do cities like Manchester and Birmingham which have risen to prominence only since the Industrial Revolution. At the other end of the scale many places rate only brief entries, being identified with only a single personality or event. To take account of these disparities major centres – around a dozen in all – have been singled out as the subject of relatively long essay entries.

To an historian, precise dates are often of no particular intrinsic significance. Nevertheless, they do help to indicate the sequence of events – much as punctuation gives order to prose – and the relative ages of people concerned. The dates of birth and death of individuals have generally been included whenever they occur and not, as is more usual, at first mention. This reflects the nature of the book, which is not designed for continuous reading: dates – if they are wanted at all – need to be found within the entry in question. For those wanting to work back from people to places there are also brief biographies of most of the more important people mentioned.

As the places listed are scattered the length and breadth of Great Britain and Ireland, I can claim direct knowledge of relatively few. I have, therefore, relied heavily on published information (see Further Reading, p. 245). Additionally, I am indebted to many people – academic staff, museum curators, regional tourist information staff, industrial liaison officers, officers of learned societies, and others – for their helpful replies to my inquiries. Nevertheless, for all errors I am myself accountable. Although some 750 places are included, this gazetteer makes no claim to be comprehensive and my choice is to some extent subjective. In compiling it, the operative word was connotation: what people, structures, inventions, or events does mention of a place conjure up? I hope that there are neither too many major omissions nor too many trivia.

This is a ramble that I have promised myself for many years. I have very much enjoyed it and hope that others will do so also. For those who wish to explore further, a short list of Further Reading is appended.

Trevor I. Williams
Oxford, June 1996

LOCATION OF PLACES MENTIONED

Almost all the places mentioned will be easily found with the help of one of the larger road atlases (2–3 miles to the inch). Where this is not so – as, for example, with some sites of archaeological interest – they have additionally been located in relation to a place that is so named.

The primary location is in accordance with current county boundaries (regions in the case of Scotland) but a word of caution is needed. Over the years – and most particularly the last twenty years – county boundaries have been redrawn, often substantially, and new names assigned. For example, in Wales what we used to know as Pembroke is now Dyfed; Humberside will not be found on older maps. Generally, this is not a problem except when information has had to be taken from old reference books: it may, for example, be difficult to assign an old Yorkshire reference with certainty to the present North, West or South Yorkshire. Within individual entries, places mentioned that have separate entries of their own are, where appropriate, distinguished by capitals.

The following abbreviations were used to distinguish the different counties and regions while the work was in preparation. Very recently (April 1996) the counties of Avon, Cleveland and Humberside were each replaced by four unitary authorities.

England

Avon	Avon	Durham	Dur.
Bedfordshire	Beds.	Essex	Esx
Berkshire	Berks.	Gloucestershire	Glos.
Buckinghamshire	Bucks.	Greater London	GL
Cambridgeshire	Cambs.	Greater Manchester	GM
Channel Islands	Ch. Is.	Hampshire	Hants.
Cheshire	Ches.	Hereford & Worcester	H & W
Cleveland	Clev.	Hertfordshire	Herts.
Cornwall	Corn.	Humberside	Humbs.
Cumbria	Cumb.	Isle of Man	IoM
Derbyshire	Derbs.	Isle of Wight	IoW
Devon	Devon	Isles of Scilly	IsS
Dorset	Dors.	Kent	Kent

Lancashire	Lancs.
Leicestershire	Leics.
Lincolnshire	Lincs.
Merseyside	Mers.
Norfolk	Nflk
Northamptonshire	Northants.
Northumberland	Northld
Nottinghamshire	Notts.
Oxfordshire	Oxon.
Shropshire	Shrops.
Somerset	Som.
Staffordshire	Staffs.
Suffolk	Sflk
Surrey	Sry
Sussex (East)	E Ssx
(West)	W Ssx
Tyne & Wear	T & W
Warwickshire	Warks.
West Midlands	W Mids
Wiltshire	Wilts.
Yorkshire (North)	N Yorks.
(South)	S Yorks.
(West)	W Yorks.

Scotland

Borders	Bdrs
Central	Cent.
Dumfries & Galloway	D & G
Fife	Fife
Grampian	Gramp.
Highland	Hghld
Lothian	Loth.
Orkney	Ork.
Shetland	Shet.
Strathclyde	Strath.
Tayside	Tays.
Western Isles	W Is.

Wales

Clwyd	Clwyd
Dyfed	Dyfed
Glamorgan (Mid)	M Glam.
Glamorgan (South)	S Glam.
(West)	W Glam.
Gwent	Gwent
Gwynedd	Gynd
Powys	Powys

Northern Ireland

Antrim	Ant.
Armagh	Arm.
Down	Down
Fermanagh	Ferm.
Londonderry	Londy
Tyrone	Tyr.

Republic of Ireland

Carlow	Carl.
Cavan	Cav.
Clare	Clare
Cork	Cork
Donegal	Dngl
Dublin	Dub.
Galway	Gal.
Kerry	Kerry
Kildare	Kild.
Kilkenny	Kilk.
Laois	Laois
Leitrim	Leit.
Limerick	Lim.
Longford	Long.
Louth	Louth
Mayo	Mayo
Meath	Meath
Monaghan	Mongh.
Offaly	Ofly
Roscommon	Rosc.
Sligo	Sligo
Tipperary	Tipp.
Waterford	Wat.
Westmeath	Wmth
Wexford	Wex.
Wicklow	Wklw

ABERCRAF, Powys

Cerrig Duon (5 miles NE) is a small, well-preserved circle of twenty stones, dating from the Bronze Age.

ABERCRAVE, W Glam.

The Dan-yr-Ogof Caves in the Upper Swansea Valley are one of the largest underground complexes in Europe. Some of the caves are open to the public and there is an associated museum devoted to cave archaeology, associated animals and the lifestyle of the cave-dwellers.

ABERDEEN, Gramp.

An important seaport and fishing centre, which in recent years has also become an important focus for the North Sea oil industry. In the harbour, the North Pier, completed in 1815, is a memorial to two great civil engineers, John Smeaton (1724–92) and Thomas Telford (1757–1834). Much of the city is built of granite, the most readily available stone, and it is often called the Granite City: quarrying is an important local industry. A less forbidding title is the Silver City by the Sea, an allusion to the way in which granite glistens when sunshine follows rain. The Aberdeen Maritime Museum, Shiprow, depicts local industries including shipbuilding, fishing and oil. Looking inland, the major industries are farming and forestry and these interests are reflected in two famous research institutes: the Rowett Institute for Research in Animal Nutrition and the Macaulay Institute for Research in Soil Science. The first Director of the Rowett was J.B. Orr (1880–1971) who, as Lord Boyd Orr, later achieved fame as the first Director of the UN Food and Agriculture Organization. The first Director of the Macaulay was Sir William Ogg FRS (1891–1979) who later became Director of the ROTHAMSTED Experimental Station. The MAFF Torry Research Station is devoted to all aspects of the fishing industry.

The University is the most northern in Britain. It was founded as King's College in 1494 and authorized by a Papal Bull – still preserved in the library – in 1495. In 1593 a rival Protestant university, Marischal College, was established only a mile away by Earl Marischal, a leader of the Reformation Party. Over the years many attempts were made to effect a union but this was not achieved until 1860, and then only after much acrimonious debate. It has a fine record of academic achievement, especially considering that its relative isolation from the other ancient Scottish universities – more than a hundred miles north of Edinburgh, Glasgow and St Andrews – made it inevitable that ambitious staff and students would tend to move nearer the centres of scholarly activity. Thus while many scientists of distinction have been associated with the University, their stay has tended to be brief.

In the 17th and 18th centuries the name Gregory had a particular significance. James Gregory FRS (1638–75), a student of Marischal, devoted himself to astronomy and mathematical optics. After travelling in Europe he was appointed to a new chair of mathematics at Aberdeen in 1669, before moving on to Edinburgh five years later. About 1663, he invented the Gregorian reflecting telescope in which elliptical mirrors replaced glass lenses, thus eliminating the coloured 'fringes' which result from light of different wave-lengths being slightly differently refracted by glass. His nephew David Gregory FRS (1659–1708) also studied at Marischal and was appointed Professor of Mathematics at Edinburgh at the age of 23. He then progressed to be Savilian Professor of Astronomy at Oxford. He is remembered as 'an astronomer who never made an observation', yet Newton – whose new philosophy Gregory enthusiastically expounded – regarded him as ' . . . in mathematics a great artist'. Another who helped to promote Newton's ideas was Colin Maclaurin FRS (1698–1746) whose *Account of Sir Isaac Newton's Philosophical Discoveries* was posthumously published in 1748. He was appointed Professor of Mathematics at Marischal in 1717, while still in his teens, and moved to Edinburgh in 1725.

Yet another member of the Gregory family was John Gregorie FRS (1724–73), grandson of James. After practising in London he was Mediciner at King's, 1756–66. Alexander Forsyth (1679–1843) was another student of King's who made a name for himself, but in a very different field. He dabbled in chemistry and mechanics and was appointed minister at Belhelvie, near Aberdeen. He is remembered as the inventor of the percussion lock for firearms, replacing the traditional flint and steel by mercury fulminate. He demonstrated his invention in the Tower of London in 1806: a memorial plaque was placed there in 1929, of which there is a replica in King's. He declined an offer of £20,000 from Napoleon to reveal the secret but was awarded a state pension. David Gill FRS (1843–1914), who also had a taste for mechanics, was born in the year that Forsyth died. He was a student at Marischal and son of a well-established clock and watchmaker in Aberdeen. He was sent to study clockmaking in Switzerland, Coventry and Clerkenwell, and then returned to succeed his father for ten years in the business. Later he made a name for himself as an astronomer and in 1879 was appointed HM Astronomer at the Cape of Good Hope.

Still in the 19th century, Aberdeen had its fair share of distinguished scientists. Undoubtedly the greatest of these – for he was a man of giant intellectual stature – was James Clerk Maxwell (1831–79), who was Professor of Natural Philosophy at Marischal 1852–6; he married a daughter of the College Principal. Later he became the first Cavendish Professor of Experimental Physics at CAMBRIDGE. He is remembered particularly as the originator of the electromagnetic theory of light. In the same general sphere of interest was Frederick Soddy FRS (1877–1956), originator of the concept of isotopes, a word which he coined: he was Professor of Physical Chemistry at Aberdeen 1914–19, before going to OXFORD. He was awarded a Nobel Prize in 1921.

In the biological sciences, too, there are plenty of noteworthy names. Robert Brown FRS (1773–1858) was a student at Marischal. He was one of the greatest of all plant taxonomists and his name is remembered in the Brownian Movement, the random movement of pollen and other very small particles suspended in liquids.

One of the first graduates of the 'new' university in 1863 was Sir David Ferrier FRS (1843–1928) who made a name for himself in London as a neurologist and pioneer investigator of brain function. His microscope is preserved in the Marischal library. Another graduate was Sir Patrick Manson FRS (1844–1922): named the Father of Tropical Medicine, he is famous for his research on insect-borne diseases, especially malaria.

One of the great medical advances of the 20th century was the introduction of insulin in 1922 for the treatment of diabetes. The two scientists most directly responsible were F.G. Banting (1891–1941) and J.J.R. Macleod (1876–1935): they shared a Nobel Prize in 1923. Macleod was very much an Aberdonian. Although born at CLUNY, Fife, he was educated at Aberdeen Grammar School and then read medicine at the University. In 1903 he moved to North America and in 1918 was appointed Professor of Physiology at Toronto University and it was there that he collaborated with Banting in their historic research. In 1926 he returned to Aberdeen as Regius Professor of Physiology.

ABERFELDY, Tays.

The two concentric stone circles of Croft Moraig lie close to the Aberfeldy–Kenmore road. Started in the Neolithic, it was altered in the Bronze Age. At Lundin Farm (2 miles NE) there is a good example of a 'four-poster' circle – four stones set at the corners of a square.

ABERFOYLE, Cent.

For more than a century Glasgow has drawn water from Loch Katrine, nearly 40 miles north, one-third of it flowing through tunnels. The engineer was J.J.LaT. Bateman (1810–89) and the work was completed in 1859. His aqueduct crosses the River Duchray south of Aberfoyle; it consists of cast-iron troughs and wrought-iron tubes.

ABERGELE, Gynd

The Iron Age promontory fort at Dinorben was defended by three lines of ramparts, of which the inner one, in stone, encircles the whole site. Two guard rooms are built into the thick walls at the entrance. The sites of some fifty round-huts can be seen within.

ABERLADY, Loth.

The Myreton Motor Museum displays a large collection of vehicles. There are a few early examples, such as an 1897 Arnold Benz and a 1902 Wolseley, but most are post–1920. All are roadworthy and frequently take part in rallies and feature in film and television shows.

There are also some bicycles from 1866, motor-cycles dating from 1903 onwards, and many motoring memorabilia such as road signs and petrol cans.

ABERNETHY, Tays.

From the 10th century many monastic sites in Ireland were dominated by round-

3

towers, serving both as a belfry and a place of refuge in time of danger. Only three, including the Abernethy Tower, are known outside Ireland. (See PEEL.)

ABERTILLERY, Gwent

The Abertillery Museum is devoted primarily to local history and accordingly features many exhibits relating to the once-flourishing local iron and coal industries.

ABERYSTWYTH, Dyfed

University town and popular resort. The university (a college of the University of Wales) is particularly known for the Welsh Plant Breeding Station, internationally famous for its work on grassland improvement. Among its distinguished graduates was Sir John Russell FRS, who eventually became Director (1912) of the ROTHAMSTED Agricultural Station.

ABRIACHAN, Hghld

Crofting, once the staple way of life in the Highlands, has declined and much of the land is now devoted to forestry and grazing. The Abriachan Croft Museum displays tools and implements used in crofting, as well as typical household equipment.

AGHAGALLON, Down

Riverdale Mill, water-powered, drives a variety of agricultural machinery, including a potato-sorter, grindstones and a milk churn.

ALBRIGHTON, Shrops.

The Aerospace Museum (RAF Cosford) displays a very large collection of military and civil transport aircraft, missiles, etc.

ALBURY, Sry

William Oughtred (1575–1660), regarded as the leading mathematician of his generation, was Rector of Albury, 1608–60. A staunch Royalist, he is reputed to have died of a transport of joy on hearing of the restoration of Charles II.

ALCESTER, Warks.

Ragley Hall, seat of the Marquis of Hertford, was built in the Palladian style in 1680. The stable block displays a collection of carriages.

ALDEBURGH, Sflk

Aldeburgh Moot Hall Museum, in a 16th-century building, is devoted to local history, with emphasis on trade and shipping. It also displays the main treasures from the Snape Ship Burial, older than the better-known Sutton Hoo Ship Burial.

ALDERSHOT, Hants.

Since 1856 Aldershot has been an important centre for the British Army and houses several museums representing various aspects of military technology. The Aldershot Military Museum (Queens Avenue) is located in the only surviving example of the original bungalow-style barracks. Exhibits include a reconstructed barrack-room and a model of the cavalry barracks. Other exhibits commemorate the early days of military flying, from balloons onward. The Army provided casual medical services from the mid-17th century but the Royal Army Medical Corps was not formally constituted until 1898. The RAMC Historical Museum (Keogh Barracks) has exhibits depicting army medical services over the last 350 years. The history of dental services are displayed in the Royal Army Dental Corps Historical Museum (Evelyn Woods Rd).

For all armies efficient transport has always been of paramount importance. The Royal Corps of Transport Regimental Museum (Buller Barracks) depicts the changing modes of transport in the British Army from the 18th century. Until the turn of the century, horses and mules were the main source of motive power and veterinary medicine had a correspondingly important role. This is illustrated in the Royal Army Veterinary Corps Museum (Galloway Rd).

ALFORD, Gramp.

The Grampian Transport Museum has a large collection of road vehicles, many with local associations, and related items. Also workshop tools. One of the most unusual exhibits is a steam-powered tricycle built by a local postman in 1895.

The Alford Valley Railway Museum is a separate unit, devoted particularly to the history of the Great North of Scotland Railway. It operates a narrow-gauge passenger service between Alford and the country parks at Murray and Haughton.

ALICE HOLT, Hants.

Once an important centre of the New Forest pottery industry. A Roman kiln has been reconstructed, and many dumps of kiln waste, now overgrown, can be seen.

ALLOA, Cent.

Alloa became industrially important as a coal port in the 17th century. To promote this the Earl of Mar constructed a primitive railway in 1768, passing in tunnels under the town; it is still possible to walk the track.

Glassworks were established in 1750 and a cone furnace still stands in Glasshouse Lane. At Alva (4 miles N) is the massive Strude Mill, dating from 1820.

ALSTON, Dur.

Killhope Lead Mine (8 miles SE on A689) has long been closed but the ruins of the original buildings, and a huge 34-ft waterwheel, stand out conspicuously on the moor.

ALTON, Hants.

Birthplace of the botanist William Curtis (1746–99), son of a tanner, and founder of *The Botanical Magazine*. Fourteen volumes, richly illustrated with the world's flowers, appeared in his lifetime. He established a private botanic garden in Lambeth, and later at Brompton, and was a founder member of the Linnaean Society in 1788.

ALUM BAY, IoW

The Museum of Clocks exhibits over two hundred clocks and watches, all in working order. Most are of conventional design, but the exhibits include an early battery-powered electric clock, a Chinese alarm clock including an elaborate device triggered by the burning of an incense stick, and a Bavarian trumpeter clock.

Alum Bay itself is famous for its multicoloured sands.

ALVINGHAM, Lincs.

Alvingham Water Mill, an 18th-century mill powered by a breastshot wheel, is now a working museum.

AMBERLEY, W Ssx

The Amberley Chalk Pits Museum is a large open-air museum with exhibits illustrating many aspects of industrial history and rural and domestic crafts. It is located in a quarry which served a local lime works 1840–1960. Many of the old buildings have been restored and put to use. A section devoted to transport includes road-rollers, both steam and diesel, and industrial narrow-gauge rolling stock.

AMBLESIDE, Cumb.

The Roman fort of Mediobogdun was built to defend the Hardknott Pass (12 miles W). Much of the outer wall and corner towers survive to a height of 3 m. Internally, the foundations of the headquarters building (*principia*) and granaries have been uncovered.

AMESBURY, Wilts.

Stonehenge, one of the greatest and most famous prehistoric monuments in Europe, is located on Salisbury Plain 2 miles W of Amesbury. It has been designated a World Heritage Site by UNESCO. A huge stone circle, it was probably commenced about 2800 BC and completed about 2000 BC. It consists of an outer ring of huge lintelled uprights and an inner horseshoe. Individual stones weigh up to 50 tons, and many were quarried at least 20 miles away. More remarkably, some smaller stones – weighing around 4 tons each – were brought from the Preseli Mountains in south-west Wales, nearly 200 miles away. Near by is Woodhenge, discovered by aerial reconnaissance in 1926. It consisted of concentric rings of wooden pillars which have long since disappeared, but the postholes have been located and indicated by concrete markers.

Immediately to the north of Woodhenge lie the Durrington Walls, a vast but now unimpressive henge monument having a maximum diameter of 520 m.

Sir Oliver Lodge (1851–1940), pioneer of wireless telegraphy, is buried in the churchyard of St Michael's, Wilsford, just south of the town.

AMLWCH, Gynd

The famous Parys Mountain copper mine on Anglesey was worked in Roman times. It then lay idle until the 1780s when, in the hands of Thomas Williams (1737–1802), it became the largest copper mine in the world, though it languished after his death. For a time, 1000 tons of copper were smelted annually at the company's works in Lancashire. Michael Faraday visited the mines during a tour in Wales in 1819 and was much impressed.

ANDOVER, Hants.

The Andover Museum (Church Close) is housed in the old Andover Grammar School. A central feature is a gallery devoted to the history of Tasker's, a local engineering firm which specialized in making agricultural machinery. The firm's original material has been considerably complemented. Other galleries depict local archaeology and geology.

Danebury Hill fort, dating from the Iron Age and covering two acres, has been extensively excavated in recent years. It controlled a considerable area and was an important grain store. Many of the artefacts recovered are displayed in the Museum of the Iron Age, an extension of Andover Museum.

ANNAGHMORE, Arm.

The Ardress House and Farmyard is a complex now belonging to the National Trust. The farmyard, built in the 19th century, displays a variety of agricultural machinery, all of local interest.

ANNALONG, Down

Annalong Cornmill, powered by water, dates from the early 19th century, and was operational until the mid-1960s. It is now restored as a museum.

ANSTRUTHER, Fife

The Scottish Fisheries Museum, housed in old buildings dating back to the 16th century, illustrates the history of the industry by means of ships, models and memorabilia. Fishing is widely construed to include whaling and commercial salmon fishing. Exhibits include a reconstruction of a fisherman's cottage as it would have appeared around the beginning of this century.

ANTONINE WALL, Cent., Strath.

In AD 122 the Romans' northern limit in Britain, and indeed in Europe, was delineated by HADRIAN'S WALL. Twenty years later it was superseded by a wall further north, ordered by the Emperor Antoninus Pius (AD 86–161) joining Carriden (Cent.) on the Forth with Old Kilpatrick (Strath.) on the Clyde. One reason, no doubt, was that apart from enclosing more territory the Antonine Wall was only half the length (37 miles) of Hadrian's (75 miles).

Strictly speaking, the Antonine Wall is not a wall at all but a rampart of turf or clay laid on a stone foundation. It followed the line of a temporary frontier established by Agricola in AD 81. Lacking enduring masonry, it is much less impressive than Hadrian's, and in many places only its general line can now be discerned. It was defended by a number of timber-built forts, long since perished above ground, perhaps twenty in all. It does not seem to have been particularly successful, for it is known to have been overrun at least twice and it was abandoned by AD 200.

A good impression of the wall can be gained at Rough Castle, near BONNYBRIDGE. Here three sides of a fort are delineated by an earth rampart: the fourth side is formed by the wall itself.

APPLEDORE, Devon

The North Devon Maritime Museum depicts – through models, paintings, photographs and in other ways – the ships that frequented the North Devon seaports from the 17th century. Other models depict aspects of shipbuilding, such as the making of ropes and sails, and local sea fishing.

ARBORFIELD, Berks.

The servicing of all kinds of military equipment has long been an important task in the British Army. Traditionally, this was done by the Royal Army Ordnance Corps but in 1942 a new body, the Corps of Royal Electrical and Mechanical Engineers (REME) was set up to take account of changing technology. The REME Museum was established in 1958 to preserve historical information relating to REME and its predecessor.

ARBROATH, Tays.

Smoking is one of the oldest methods of preserving fish and it has been practised here for centuries: Arbroath Smokies are world famous. There are still a number of working smokeries and a small museum is housed in the signal tower built for the Bell Rock Lighthouse in 1813.

ARKLOW, Wklw

In the middle of the 19th century, more than 150 sailing ships were owned in Arklow and they traded as far afield as North and South America and to many European ports. The Arklow Maritime Museum (St Mary's Rd) depicts many of these vessels in models and paintings. Also on view are sections of the transatlantic cables laid by the *Great Eastern* (1857/8). The ship was commanded by Captain Halkin, a local man.

ARMAGH, Arm.

The Observatory on College Hill has a planetarium and a small collection of scientific instruments. James Ussher (1581–1656), Archbishop of Armagh, may be regarded as a pioneer cosmologist. Although his conclusions were based on Holy Writ, rather than scientific evidence, his calculation that the Creation occurred in 4004 BC was accepted for many generations.

ARNOL, Lewis, W Is.

Black House – a typical Lewis farmhouse – and farm have been preserved, suitably furnished and equipped, in Arnol village.

ARRAN, Strath.

The Isle of Arran is rich in archaeological remains. These include a complex of six stone circles and standing stones on Machrie Moor and, at Auchegallon, a Bronze Age cairn situated within a ring of stones.

ARUNDEL, W Ssx

The Arundel Museum and Heritage Centre, in the centre of the town, is largely devoted to local history. Additionally, there is an interesting historic collection of weights and scales.

ASHBURTON, Devon

Grimspound (6 miles NW) is a well-preserved example of the many Bronze Age settlements on Dartmoor. The outer wall, 3 m thick, encloses an area of nearly an acre, and in places still stands to a height of 1.5 m. Within, the sites of some thirty hut-circles can be seen.

ASHBY-DE-LA-ZOUCH, Leics.

Birthplace of Sir Frank Dyson FRS (1868–1939), Astronomer Royal 1910–33. He made important observations of solar eclipses, in particular that of 1919 in which Einstein's predicted gravitational deflection of light passing the Sun was observed.

ASHFORD, Kent

Birthplace of John Wallis (1616–1703) mathematician and founder member of the Royal Society (see OXFORD). His famous *Arithmetica Infinitorum* (1655) is mainly concerned with the use of mathematical series with geometrical problems. He devised a famous series to evaluate π and invented the symbol ∞ for infinity.

ASHFORD, Wklw

The Mount Usher Museum displays a collection of carriages made in Dublin.

ASHTON-UNDER-LYNE, GM

The Portland Basin Industrial Heritage Centre – housed in an old warehouse, with an adjacent water-wheel – depicts two centuries of Tameside industrial and social history.

ASHWELL, Herts.

Ashwell Village Museum (Swan Street) is located in a fine example of a 16th-century timber-framed house, once belonging to Westminster Abbey.

ASTON MUNSLOW, Shrops.

The Whitehouse Museum of Buildings is a 6-acre complex where the evolution of building methods since 1043 is displayed. At its centre is a homestead which was occupied by the same family from 1332 until 1946 and the associated buildings are ones which were put up from time to time as needed by the occupants. They include a corn-mill, cider press, dovecote, dairy and coach house.

AUCHINLECK, Strath.

As its name implies, the Auchinleck Boswell Museum and Mausoleum is primarily devoted to James Boswell, the biographer of Samuel Johnson, the great lexicographer. One section, however, is devoted to William Murdock (1754–1839), pioneer of gas lighting, who was born nearby at Old Cumnock. (See BIRMINGHAM, REDRUTH.)

AUCHTERARDER, Tays.

The Strathallan Aircraft Museum (Strathallan Airfield) displays a collection of aircraft, mostly of historic interest. Marques on view include a Lancaster, a Shackleton and a Comet airliner. There is also a collection of aero-engines, some dating from the First World War. A unique exhibit is the original Rolls Royce VTOL (Flying Bedstead) aircraft.

AULT HUCKNALL, Derbs.

Thomas Hobbes (1588–1679), political and natural philosopher, is buried in the church of St John the Baptist.

AUST, Glos.

The River Severn is here spanned by the Severn Suspension Bridge, completed in 1966, carrying the M40 motorway. It has a span of 988 m. A second bridge was completed some miles downstream in 1996. The cliffs adjoining the estuary here are very popular with fossil hunters.

AVEBURY, Wilts.

The huge stone circle-henge here is one of the great prehistoric monuments of Europe, dating probably from the late 3rd millennium BC. The outer of three circles contains nearly a hundred sarsen boulders, and encloses two smaller rings. With Stonehenge it has been designated a World Heritage Site by UNESCO.

Silbury Hill is 40 m high, covers a ground area of 2 ha, and dominates the local landscape. It is the largest prehistoric mound in Europe, and was probably built around 2000 BC. It is generally regarded as a huge Neolithic barrow, but no human remains have been found despite several excavations. It has been estimated that some 20 million man-hours would have been needed to shift the 14,000 tons of soil required for its construction. The nearby West Kennett Long Barrow is 100 m long, and access has been made to a large internal burial chamber.

AXBRIDGE, Som.

King John's Hunting Lodge (NT) has exhibits depicting local history, especially that of the wool and cloth trade. The Lodge itself has nothing to do with King John, but architecturally it is a fine example of an early 16th-century merchant's house.

AXMINSTER, Devon

The town is world-famous as a centre of the carpet industry, the history of which is portrayed in the Axminster Museum (Church St). The setting is unusual: a Victorian police station and courtroom, complete with cells.

AYLESFORD, Kent

Kits Coty House is the remains of a burial chamber; a massive capstone is supported by three uprights.

AYR, Strath.

Birthplace of John Loudon McAdam (1756–1836), the great road builder, whose name is remembered in the 'macadamized' road surface. There is a memorial to him in the town. (See BRISTOL)

AYSGARTH, N Yorks.

The collection of horse-drawn vehicles in the Yorkshire Carriage Museum is one of the most comprehensive in Britain, ranging from milk floats to goat carts. It is housed in Yore Mill, a water-powered textile mill built in 1784. The vehicles are all roadworthy and many appear from time to time on film and television. There is also a small collection of related small articles, such as harness.

B

BAGINTON, Warks.

The Lunt Roman Fort was used primarily for cavalry. Much restoration work has been done and a clear picture has emerged of the fort as it must have appeared in the 1st century. Recovered artefacts are displayed in an old granary.

BAILLIEBOROUGH, Meath

There are two barrow cemeteries on Mt Cairnbone. The best preserved is Cairn T, on Cairnbone East. It is nearly 40 m in diameter and the stones in the burial chamber are incised.

BAKEWELL, Derbs.

Chatsworth House (National Trust) was built for the 1st Duke of Devonshire between 1687 and 1707. The contents include a collection of scientific instruments, many associated with the chemist and natural philosopher Henry Cavendish (1731–1810), grandson of the second duke. His name is commemorated in the famous Cavendish Laboratory, CAMBRIDGE. The conservatories were built by Joseph Paxton (1803–68) who used them as a model for his huge Crystal Palace, built for the Great Exhibition of 1851.

The Old House Museum displays some of the early textile machinery of Sir Richard Arkwright (1732–92).

BALLINAMORE, Leit.

Ballinamore is close to a number of sites of archaeological interest and these are represented in the general exhibition in Ballinamore Local Museum. Additionally, there are displays of household equipment, craftsmen's tools and agricultural machinery from the late 19th century.

BALLINDALLOCH, Gramp.

Until 1823 the making of whisky in Scotland was illegal. In that year the government decided that the best way of curtailing the illicit trade was to legitimize it. This led to the founding of many now-famous distilleries and today whisky is an important export as well as a major source of revenue. Traditional methods of production are well-illustrated at the Glenfarclas Distillery Museum.

BALLYCOPELAND, Down

Ballycopeland Windmill is a tower mill, with adjacent kiln and miller's house, dating from the late 18th century. It was operational until 1915 and has been restored to full working order.

BALLYDUGAN, Down

Here, by Lake Ballydugan, are the remains of an eight-storey flour mill, and an engine-house and smokestack dating from 1792.

BALLYGLASS, Mayo

The fine megalithic court cairn here has been built over the ruins of an earlier house.

BALLYNAHOWNE, Gal.

Birthplace of the chemist William MacNeven (1763–1841). After studying medicine in Vienna, he emigrated to America in 1805, where he was Professor of both chemistry and medicine in the College of Physicians and Surgeons in New York, before moving to Rutgers College (now University). He has been described as 'the Father of American Chemistry'.

BAMBURGH, Northld

Bamburgh Castle, once the home of the Kings of Northumbria, was restored in the late 19th century. It displays a fine collection of armour, some on loan from the Tower of London.

BANBURY, Oxon.

Market town: an important centre of the aluminium industry (Alcan). Birthplace of Sir Alan Hodgkin (1914), Nobel Laureate 1963, noted for his research on the mechanism of transmission of nervous impulses. (See CAMBRIDGE.)

BANFF, Gramp.

The Banff Museum has a small collection of scientific instruments, some associated with the astronomer and instrument maker James Ferguson (1710–76) who was born nearby at Rothiemay and worked in London from 1743.

BANGOR, Gynd

The famous suspension bridge over the Menai Straits is a memorial to the great civil engineer Thomas Telford (1757–1834). It stands 100 ft above the tideway and has a span of 579 ft. Opened in 1826, it was an essential part of his general improvement of the London-Holyhead road. The nearby Britannia Bridge carrying the railway was built by Robert Stephenson (1803–59) and William Fairbairn (1789–1874). It had a remarkable structure, consisting of two cast-iron tubes, each with a span of 460 ft. Two million rivets were used in its construction. In 1970, it was severely damaged by fire and was rebuilt ten years later with a continuous beam, with a road for motor traffic above.

The Museum of Welsh Antiquities (Ffordd Gwynedd) has displays illustrating local crafts and industry, including the mining and working of slate, fishing and textiles.

BARMOUTH, Gynd

Barmouth, now mainly a holiday resort, was until the turn of the last century a flourishing seaport and the Barmouth lifeboat was famous. The RNLI Museum commemorates this and illustrates the technical aspects of lifeboat construction and operation.

BARNET, GL

Birthplace of John Hadley FRS (1682–1744), son of a High Sheriff of Hertfordshire. He established himself as a very successful instrument maker. In 1730, he designed a reflecting quadrant, predecessor of the modern sextant, which became very popular among navigators because of its convenience in use. To this, he added in 1732 a bubble-level for use when the horizon was not visible.

BARNSLEY, S Yorks.

Worsborough Mill Museum (3 miles S) is located in a 17th-century water mill and a 19th-century (1840) steam-powered mill.

BARROW-IN-FURNESS, Cumb.

Barrow first rose to prominence in the 19th century as a port for shipping local iron ore to south Wales and the Midlands, but from 1859 steel-making was begun and shipbuilding gradually became the dominant industry. From the early 20th century the construction of submarines became a speciality.

These aspects of the town's history are depicted in the Furness Museum (Ramsden Sq), which includes a number of ship models.

BARTON, Lancs.

In 1759 the Duke of Bridgewater was empowered to build a canal to carry coal to Manchester from his pits at Worsley and engaged James Brindley (1716–72), a semi-literate millwright, to carry out the work, which was completed in 1761. The most remarkable feature of the canal was the Barton Aqueduct, carrying the canal over the River Irwell. This was the first British aqueduct capable of carrying boat traffic. It was replaced by a unique swinging aqueduct, still in use.

The face of one of Brindley's arches has been preserved alongside the Barton–Eccles road.

BATH, Avon

Fashionable West Country spa famous for its Roman Baths (museum in Abbey churchyard). It has been designated a World Heritage Site by UNESCO. Home of William Herschel (1738–1822). He supported himself by teaching music and in 1766 was appointed organist to the Octagon Chapel. However, aided by his daughter Caroline, he became keenly interested in astronomy and learnt to construct his own instruments. In 1781 he observed what he thought was a comet and so described it to the Bath Philosophical Society, but it proved to be a new planet, Uranus. This brought Herschel international fame. George III appointed him Court Astronomer and awarded him a pension of £200, supplemented by £50 for Caroline, conditional on his moving to the neighbourhood of Windsor and occasionally showing the heavens to the royal family. Herschel House and Museum is at 19 New King Street.

John Wood (1705?–54) settled in Bath in 1727 and achieved fame as an architect in the Palladian style. He designed the North and South Parades, Queen's Square, the Circus, and other elegant buildings. He is commonly known as 'Wood of Bath'. William Oliver (1695–1764), physician to Bath Mineral Water Hospital 1740–61, is famous for the Bath Oliver biscuit.

Birthplace of Richard Lovell Edgeworth (1744–1817), inventor and educationist. His three-volume *Practical Education*, written with his daughter Maria, the novelist, was a great success. From 1766–82 he was a popular member of the Lunar Society of BIRMINGHAM. The tomb of Robert Malthus (1766–1834), author of *Essay on the Principle of Population*, is in the Abbey.

The Royal Photographic Society's National Centre of Photography is in Milsom Street. It houses a fine collection of cameras and photographic equipment, totalling some six thousand items. There is a Museum of Bookbinding in Manvers Street. The Bath Industrial Heritage Centre, Camden Works, houses the machinery and tools used by J.B. Bowler, brass founders, in their engineering business from 1872 to 1969.

The Kennet and Avon Canal runs south of the city and this section is noteworthy for the cast-iron Dundas Aqueduct built by John Rennie (1761–1821). Also on the canal, at Claverton (2 miles E) are water-powered beam pumps, now restored. Since Domesday there has been a water-powered corn mill at Priston (5 miles SW) but the present mill dates largely from the 18th century.

BATHGATE, Loth.

Cairnpapple Hill was a complex ritualistic site from Neolithic times up to the Iron Age, and successive stages of development can be discerned at the site. Many urn cremations have been unearthed.

BATLEY, W Yorks.

Industrial centre, near Leeds. Here, woollen shoddy for respinning was first made in the 1830s by breaking down rags. The machinery was invented by Benjamin Parr. It is said that the alternative name of mungo arose because Parr insisted that all the rags 'mun go' through the carding machinery.

BATTLESBRIDGE, Esx

Tide mills are a special form of water mill. The rising tide flows into a reservoir, from which it is released by a sluice as needed to supply the water wheel. This example on the River Crouch dates from the late 18th century. There is an associated kiln for drying grain, and a granary.

BEAMISH, Dur.

The North of England Open Air Museum displays local life as it appeared around 1900. It includes a colliery village, home farm and railway station. It also preserves an important early steel furnace, using the cementation process, at Hamsterley. Nearby, at Causey, is the Causey Arch, reputedly the oldest surviving railway bridge in Britain.

With a span of 32 m it was built in 1727 to cross the Tanfield Beck at a height of 24 m. The arch is now open only to foot passengers.

BEARSTED, Kent

Birthplace of Robert Fludd (1574–1637), physician and philosopher. Deeply imbued with the Rosicrucian philosophy he yet supported the scientific discoveries of his day. He thus very much epitomizes the transition from magic to science.

BEAULIEU, Hants.

The National Motor Museum has one of the world's finest collections of motor vehicles – including motor-cycles – dating from 1894. The 250 exhibits include many famous marques.

The 2nd Duke of Montagu had visions of turning Buckler's Hard into a major port, but eventually it emerged as a small shipbuilding village, and this is reflected in exhibits in the Buckler's Hard Maritime Museum. Many naval wooden-walls were built here between 1745 and 1822, including Nelson's *Agamemnon*.

BEAUMARIS, Gynd

Beaumaris Castle, at one time accessible by sea, was the last to be built in Wales by Edward I. Apart from its intrinsic interest, the Chapel Tower displays an exhibition illustrating the technical aspects of castle-building in the 13th century.

The adjacent Beaumaris Gaol Museum shows how the prison appeared when it was closed in 1878. It contains the only human treadmill in Britain still occupying its original site.

The scope of the Museum of Childhood is wider than its title suggests. Apart from toys in the traditional sense, exhibits include mechanical musical devices (musical boxes, polyphons, etc.), magic lanterns, cameras and early wireless sets.

BECCLES, Sflk

Beccles and District Museum (Newgate) displays primarily material relating to the history of the town, with particular emphasis on two local industries – printing and agricultural machinery.

Historically, Beccles is remembered for having struck its own coinage in the 17th century. Exhibits include the dies from which the rare 1670 Beccles farthing was struck, with a specimen of the coin itself.

BEDDGELERT, Gynd

The Sygun Copper Mine possibly dates from Roman times but large-scale working began only about 1830. Profitability was short-lived, however, and it changed hands several times during the 19th century. Production finally ceased in 1903 but parts of the underground workings have been restored and are open to visitors.

Technologically, the main interest of Sygun is that it was one of the first mines to adopt the Ore Flotation Process invented by Alexander Elmore, who bought the mine in 1897. Previously, mined ore had been crushed and sorted mechanically. In the flotation process, crushed ore was mixed with oil and water and the mixture left to settle. Copper sulphide particles attached themselves to the oil which floated to the top, leaving the waste to sink to the bottom. The process was quickly adopted world-wide, but it failed to restore Sygun itself to profitability.

BEDFORD, Beds.

The Bedford Museum has small collections of scientific instruments, as well as clocks and medical items.

At Old Warden aerodrome (7 miles SE) is the Shuttleworth collection of aviation history which features many flyable aircraft, ranging from a 1909 Blériot monoplane to Spitfires.

Thomas Tompion (1639–1713), 'the father of English clockmaking' was born at Northill.

BEKESBOURNE, Kent

Birthplace of Stephen Hales FRS (1677–1761), remembered particularly for his elaborate experiments, some lasting many years, on the absorption of water and air by growing plants. He summarized his finding in his *Vegetable Staticks* (1727). He also studied the circulation of the blood in animals and was probably the first to measure blood pressure. In 1709, he became perpetual curate of Teddington, refusing preferment lest it should interfere with his experiments.

BELFAST, Down

Birthplace and lifelong home of Thomas Andrews FRS (1813–85) appointed Professor of Chemistry in the newly founded Queen's College (now University) in 1849. Famous for his research on the continuity of the gaseous and liquid states, and the concept of critical temperature, which paved the way for the liquefaction of the so-called permanent gases and the creation of a great new industry. John Joly FRS (1857–1933), geologist and physicist, was born at Holywood, and was appointed Professor of Geology in DUBLIN in 1897. Home, 1867–95, of the very successful veterinary surgeon J.B. Dunlop (1840–1921) whose practice, one of the largest in Ireland, was in Gloucester Street. He is better known, however, as the inventor and pioneer of the pneumatic rubber tyre, first used for bicycles. The small company he founded ultimately grew into the giant Dunlop Rubber Company. Belfast was the birthplace of Osborne Reynolds FRS (1842–1912), pioneer of the study of liquid flow and scientific ship design. Ernest Macbride (1866–1940), famous for research in comparative embryology, was also a Belfast man. He studied at Queen's University before going on to Cambridge. He was Professor of Zoology in Imperial College, London, 1904–34.

Here in 1876 William Smiles (1846–1904) – son of Samuel Smiles the great social reformer – set up the small Belfast Ropework Co. Among the original shareholders were Edward Harland (1831–95) and Gustav Wolff (1834–1913) who had founded the great shipbuilding company in 1860. For a time the ropeworks faltered, though its fortunes were somewhat improved by persuading the great French tight-rope walker, Blondin, to use its product. When Smiles died in 1904 the company owned the biggest ropeworks in the world, covering 40 acres and supplying thirty thousand customers world-wide.

The Ulster Folk and Transport Museum exhibits include steam locomotives and rolling stock, trams, bicycles, motor-cars and commercial vehicles.

The Botanic Garden in Stranmillis Road is adjacent to the Queen's University. The garden supplies the university's biologically orientated departments and is also a popular public park. The Palm House dates from 1839. The Ulster Museum is located in its grounds: it is largely devoted to Irish archaeology and local history but also displays some linen textile machinery.

BELLEEK, Ferm.

Home of the famous Belleek Pottery, founded in 1857.

BELPER, Derbs.

Here in 1728 Jedediah Strutt (1726–97), partner of Richard Arkwright (1732–92), built a spinning mill powered by a water-wheel 7 m in diameter. This North Mill was badly damaged by fire in 1803 but it was rebuilt in 1813 and still stands. Like many industrialists of his day, Strutt associated his factories with well-planned model villages for his employees.

At Morley Park, Heage (3 miles N) are the remains of late 18th-century blast furnaces, standing in splendid isolation in open country.

BELTRING, Kent

Among the traditional crops of Kent – 'The Garden of England' – are hops. The Whitbread Hop Farm provides a substantial part of the hops required for its own brewery. On the site six Victorian oast-houses have been preserved, of which three serve as a museum. Apart from exhibits relating specifically to hops there are others illustrating horses and harness, agricultural machinery and tools, and other rural crafts.

BEMBRIDGE, IoW

The exhibits in the Bembridge Maritime Museum include navigational instruments, ship models and early diving equipment.

BENSON, Oxon.

The Benson Veteran Cycle Museum (Brook Street) displays a collection of nearly five hundred bicycles made between 1818 and 1930.

BERKELEY, Glos.

Home of Edward Jenner (1749–1823), pioneer of vaccination. His father was vicar of Berkeley, and he himself practised medicine there all his life. He is buried in the churchyard of St Mary the Virgin and a window in his memory was installed there in 1873. His home, The Chantry, houses the Jenner Museum.

Jenner was impressed by the local tradition that infection with cowpox conferred immunity to smallpox and he developed this into a systematic method of vaccination, with dramatic effect on the mortality rate of the disease. Parliament made him grants totalling £30,000. President Thomas Jefferson wrote (over-optimistically as it turned out), 'Future generations will know by history only that the loathsome smallpox existed and by you has been extirpated'. A statue by R.W. Sievier is in GLOUCESTER Cathedral.

The Berkeley Nuclear Power Station, dominating the landscape, was the first such station to reach the end of its useful life. It was shut down in 1989 in preparation for decommissioning.

BERSHAM, Clwyd

Site of the works of one of the principal 18th-century ironmasters, John Wilkinson (1728–1808). It was here that he developed his famous boring engine, originally for the boring of cannon but subsequently crucial for making the very large cylinders required for Boulton and Watt steam engines. Recently, 135 ft of a tramway with wooden rails has been uncovered. It includes a primitive points system enabling trucks to be diverted from one track to another.

The Bersham Industrial Heritage Centre has exhibits depicting the smelting and working of iron, lead mining and other local industries.

BERWICK-UPON-TWEED, Northld

At Loan End (7 miles W) is an early suspension bridge, built in 1822 by Samuel Brown (1776–1852).

The Lindisfarne Wine and Spirit Museum (Palace Green) is located in an old barrack room of the King's Own Scottish Borderers. It displays an extensive collection of equipment used in the wine and spirits industry, including instruments, such as hydrometers.

BETWS-Y-COED, Clwyd

Village on the River Conwy best known for its beautiful Swallow Falls. Technologically it is interesting for the iron bridge, cast in 1815, which carries the London-Holyhead road across the river.

Home of the Conwy Valley Railway Museum, located in the old goods yard of the local railway station. In addition to a display of standard-gauge rolling stock it includes a 7¼-inch steam railway running on nearly a mile of track.

BEVERLEY, Humb.

The Museum of Army Transport displays British Army transport for the movement of troops, equipment and supplies. Some seventy vehicles are on view, including a mobile field workshop and, in the open, a Blackburn Beverley heavy transport aircraft.

BEWDLEY, H & W

Once a busy port on the River Severn and a centre for the brass industry. The Bewdley Museum has displays of brass founding, rope-making, and other local crafts. The stone bridge over the river was built by Thomas Telford (1757–1834).

BIBURY, Glos.

The Arlington Mill Museum displays agricultural machinery and implements, a weaving loom and domestic equipment.

BIGGAR, Strath.

The Biggar Gas Light Company was founded in 1839 and was typical of hundreds of small companies established in the 19th century. It was one of the last to be closed in Scotland as a consequence of the national conversion to natural gas 1967–77. It survives as a museum administered by the National Museums of Scotland.

BIGNOR, W Ssx

The Bignor Roman Villa, discovered in the 1820s, is an archaeological site of particular interest. The mosaics on display are among the finest in Britain. The small museum displays artefacts uncovered during the excavation of the site.

BILLERICAY, Esx

The Barleylands Farm Museum houses a large collection of agricultural equipment dating back to the 16th century. It includes many tractors and some steam-ploughing engines.

The Cater Museum (High Street) is located in a house once belonging to a saddler and harness-maker and is primarily devoted to local history. There is, however, a large collection of model fire-engines, from very early to recent models.

BILSTON, W Mids

The chief local industries are iron and stone, but the town is also remembered as one of the main centres of japanned ware. Examples of this are displayed in the Bilston Art Gallery and Museum (Mount Pleasant).

BINGLEY, W Yorks.

Famous for its flight of five locks on the Leeds and Liverpool Canal. This, one of Britain's most important inland waterways, was completed in 1816 but the Bingley section was completed in 1774, as a memorial plaque commemorates.

BIRCHINGTON, Kent

Adjacent to Quex House is the Powell-Cotton Museum, housing a diverse collection amassed in Africa and Asia by Major P.H.G. Powell-Cotton (1866–1940) an avid collector and naturalist. The main exhibits are animals from all parts of the world, most of them set up by the eminent taxidermist Rowland Ward. (See also TRING).

BIRKENHEAD, Mers.

Birkenhead, now linked to LIVERPOOL by the Mersey Tunnel, has a long independent history. It was once an important centre for the production of fine ceramics, such as that of the Della Robbia Co. (1894–1901), and examples of these are displayed in the Williamson Art Gallery and Museum (Slatey Road).

BIRMINGHAM, W Mids

Although noted in the Domesday Book, Birmingham was not a town of any great consequence until the 16th century when Leland in his *Itinerarium* (1538) noted 'there be many smithes in the towne that use to make knives and all manner of cuttynge tools . . . and a great many naylors, so that a great part of the towne is mayntayned by smithes.' This activity was not fortuitous but resulted from the ready local availability of iron and coal. In 1816, in Jane Austen's *Emma*, Mrs Elton took a poor view of its future: 'One has no great hopes for Birmingham. I always say there is something direful in the sound.' Her misgivings were ill-founded: in the course of the 19th century the town grew to be one of the world's great centres for metal manufactures of all kinds and the centre of a huge industrial conurbation. It also became a great cultural centre. The famous Birmingham Musical Festival – particularly associated with Handel, Mendelssohn, Gounod and Elgar – was held triennially from 1768 until the First World War. The artists David Cox and Edward Burne-Jones were Birmingham men. Historians of science will think of the city in the context of the Lunar Society, which flourished in the second half of the 18th century. Its name has an interesting origin: it met monthly, on the afternoon of the Monday nearest full moon so that its members could more easily make their way home through the ill-lit streets. Though quite informal and small in numbers – never more than fourteen – it was the most prestigious learned society outside London. Its members included Matthew Boulton and James Watt, the great pioneers of steam-engine manufacture; the physician William Withering, famous for introducing digitalis into medicine for the treatment of dropsy; the botanist Erasmus Darwin, grandfather of Charles Darwin, proponent of evolutionary theory; Josiah Wedgwood, the potter; the radical Joseph Priestley, often called 'the father of modern chemistry'; and Richard Lovell Edgeworth, inventor of many ingenious devices and father of the novelist Maria Edgeworth.

Birmingham's industrial history is epitomized in the Museum of Science and Industry (Newhall Street). Appropriately, it is lodged in an old electro-plating factory, for this has long been an important local industry. It has extensive collections of general scientific and technological interest, including instruments, machine tools, steam engines and a letter-copying machine devised by James Watt. Steam locomotives are displayed at the Birmingham Railway Museum at Tyseley. Sarehole Mill, Hall Green, is an 18th-century watermill used variously for metal working and grinding corn. It is now a milling museum. Birmingham is famous, too, for the manufacture of small metal objects, such as buckles and buttons, and there is an exhibition of these in the Museum and Art Gallery (Chamberlain Square). The Royal Brierley Museum exhibits many fine examples of hand-made glass and crystal, another important local industry. The firm of W. & T. Avery has been famous in scale-making since the 18th century. The Avery Historical Museum at Smethwick is devoted to weighing machines of all kinds.

The first of the great Birmingham manufacturing engineers was Matthew Boulton (1728–1809) who lived his whole life there. From small beginnings, he built a large factory at Soho, making mostly small items. Then, in 1775, he went into partnership with James Watt (1736–1819) – whose statue stands in Chamberlain Square – to develop the latter's improved steam engines; with a 25-year extension of the patent, the firm of Boulton and Watt manufactured the most important prime movers of the Industrial Revolution. Ironically, steam found relatively little industrial application in Birmingham until the second half of the 19th century. Up to that time the main products were small items requiring only simple hand-operated machines. The firm had many other industrial interests: in particular it pioneered – through its chief engineer William Murdock (1754–1839) – the manufacture of coal-gas for lighting. In 1802, the Soho Works were illuminated by gas lamps to celebrate the Peace of Amiens. The Soho Manufactory was demolished in the 1860s but the foundry survives and houses the Avery Historical Museum. Boulton, Watt and Murdock are united in death as in life: all three are buried at St Mary's Church, Handsworth.

At the other end of the scale, a new metal-based industry appeared in 1821. In that year, Joseph Gillott (1799–1873) who began life in the Sheffield cutlery industry, set himself up as a manufacturer of steel pen-nibs using mass-production methods, in place of the existing technique, introduced in the 1780s, of making each nib individually using shears and file. By 1859, his large works at Newhall Hill was producing more than 5 tons of nibs per week. He amassed a huge fortune which he devoted partly to philanthropy and partly to building up a fine collection of paintings – especially by Etty and Turner – which realized £170,000 at his death. Another nib-maker, though one with many other industrial interests, was Sir Josiah Mason (1795–1881). He, too, was a great philanthropist and his greatest memorial was Mason College, founded in 1875, which has since been incorporated into Birmingham University. Its origins go back, however, to 1828 when Queen's College was founded as a school of medicine.

Among Mason's business associates was George Elkington (1801–65), pioneer of the new technique of electro-plating. In 1841 he built a large works in Newhall Street, using a cyanide process invented by a local surgeon, John Wright. He was noted for his silver-plating of delicate objects such as leaves, flowers and even spiders' webs. With Mason he established a copper-smelting business in south Wales: he also developed a process for water-proofing fabric with rubber, and sold the patent to Charles Macintosh (1766–1843) of MANCHESTER. Also associated with Elkington in the electro-plating business was another Birmingham man, Alexander Parkes (1813–90) who became known as 'the Nestor of electrometallurgy'. He also developed celluloid (cellulose nitrate) for moulding small articles, and thus has a claim to be recognized as a founder of the plastics industry.

In the main, the Industrial Revolution was brought about by men with little or no formal education but a keen awareness of the significance of new scientific discoveries and the energy and enthusiasm to make them the basis of great new industrial enterprises. Boulton was a buckle-maker's son; Watt was the son of a carpenter; Wedgwood began work at the age of nine; James Brindley, the great canal builder, was never more than semi-literate. It has been said, with reason, that science owed more to the steam-engine than the steam-engine did to science. During the 19th century, however, professional qualifications became increasingly important.

This change is exemplified by the motor-car industry, of which Birmingham has from its beginning been an important centre. Herbert Austin (1866–1941, Lord Austin of Longbridge) made his first car in 1895 and ten years later established the Austin Motor Co. in a disused printing works at Longbridge, which rapidly grew to be a huge manufacturing complex. Yet his early ambition was to be an architect, and he embarked on car manufacture only after a short apprenticeship at his uncle's engineering works in Australia. The career of F.W. Lanchester FRS (1868–1946) was very different. He trained as an engineer at the National School of Science, South Kensington, and then gained practical experience of internal combustion engines by working with a

Birmingham manufacturer of gas-engines at Saltley. In 1894 he set up his own works at Sparkbrook to manufacture motor-cars on strictly scientific principles. In those days, one, or at most, two-cylindered engines were in vogue but the unbalanced forces caused much noise and vibration. Lanchester's highly original engine had two horizontally opposed cylinders between which were two crankshafts geared together and turning in opposite directions. This engine was very nearly vibrationless. Also at Sparkbrook was located the Royal Small Arms Factory which in 1906 was amalgamated with the existing Birmingham Small Arms Co. (BSA), famous in the motor-cycle world.

The name of John Baskerville (1706–75) is famous wherever fine printing is prized. He was for a time footman in a household at King's Norton and then in 1740 set up a small but very profitable works at 22 Moor Street to make japanned goods. He also taught calligraphy and about ten years later turned his attention to type-founding, with meticulous attention to the design of the characters. In 1758, Oxford University commissioned him to provide a complete alphabet of Greek types for its forthcoming Greek New Testament. In the same year he was appointed Printer to the Cambridge University Press. In his later years he lived in a mansion – subsequently called Baskerville House – at the lower end of Broad Street. He was buried in a brick windmill built in the garden of his house, but he was not destined to rest easily. The house was gutted in the riots of 1791. In 1820 workmen uncovered the coffin but it was reburied. Six years later building operations again exposed it and for a time it was displayed in a shop in Monmouth Street. After further vicissitudes it was moved to the crypt of Birmingham Church of England Cemetery, Warstone Lane. This was demolished in 1955 but although the crypt survived it was empty. It is said that around 1830 'a surgical gentleman took a cast of the head' but of this there is no trace. Alas, poor Baskerville: you deserved a better fate than this!

Although the city is particularly identified with metal-working industries, this is by no means its only interest. The well-known Quaker firm of Cadbury – founded by George Cadbury in 1824, originally to sell tea and coffee – began to manufacture chocolate in Bridge Street in 1831. It was not a particularly successful venture and when he handed it over to his two sons, Richard and George, in 1861 it was near collapse. It was saved by what would now be termed a technological breakthrough. In 1866, the firm launched Cadbury's Cocoa Essence as a novel product. Hitherto, all cocoa products had required various additives to mask the unpleasant taste of cocoa butter, which could not be removed from the raw cacao bean. George Cadbury learnt that the Dutch firm Van Houten had developed a machine press to expel the butter and went to Holland to buy one. Coinciding with growing public concern about adulteration of food, the new product ('Absolutely pure: therefore the best') was a highly profitable success. The firm outgrew the Bridge Street factory and a 15-acre site was acquired on the Bourn stream a few miles south-west of the city. Bournville – 'the Worcestershire Eden' – was opened in 1879 as a model town to provide not only good working conditions but also a range of recreational facilities (cf. NEW LANARK, PORT SUNLIGHT, STYAL). By the end of the century, Cadbury's employed over 3000 people and annual sales exceeded £1 million: very soon the company overtook their great rival, Fry's of BRISTOL, another great Quaker business.

Among scientists particularly identified with Birmingham was the chemist Joseph Priestley (1733–1804) (see also CALNE, LEEDS). He lived there 1780–91 as minister of the New Meeting House: his statue, too, stands in Chamberlain Square. He soon became closely involved with the Lunar Society and Matthew Boulton helped to provide research facilities. Unfortunately, his genius as a chemist was not matched by his political discretion and a long stream of publications supporting the French Revolution and attacking the Established Church brought retribution. On Bastille Day, 1791, organized Church-and-King mobs burnt down Dissenting chapels in the city and the houses of some of their ministers, including Priestley's on Fair Hill. He moved to London but there too provoked hostility: in 1794 he emigrated to America.

In 1900, the University of Birmingham was the first civic university to receive a full charter and Sir Oliver Lodge FRS (1851–1940) was appointed the first Principal. He came with an international reputation for his researches on electromagnetism and pioneer experiments on radio communication. Less well respected by the scientific establishment were his investigations of psychic phenomena: he served twice as President of the Society for Psychical Research.

Over the years, the University, at Edgbaston, has grown in stature – particularly in its departments of science and engineering – and is now of international standing. In 1937, Sir Walter Haworth (1883–1950) was awarded a Nobel Prize for his research on carbohydrate chemistry. In 1939 a crucially important invention was made by H.A.H. Boot and J.T. Randall in the University's department of physics. This was the cavity magnetron, a device which generates powerful radar pulses of shorter wavelength than those previously available. It played a major role in the successful outcome of the Battle of Britain in 1940. In the post-war years, the same radiation became the basis of the microwave oven. More recently (1966) the city gained a second university, the University of Aston in Birmingham.

Among other scientists closely identified with the city was Francis Galton FRS (1822–1911) the founder of eugenics (a word he coined in 1883), who was born there. His intellectual precocity was legendary: he is reputed to have had an IQ of 200. Certainly his genes were favourable: Erasmus Darwin was his grandfather and Charles Darwin his cousin. Birmingham was also the birthplace of Sir William Napier Shaw FRS (1854–1945) who was educated at King Edward's School: he was Director of the Meteorological Office 1905–20.

BIRR, Ofly

William Parsons (3rd Earl of Rosse, 1800–67) was born at Birr Castle, the family seat. He was a keen amateur astronomer and devoted many years of his life to building his own telescopes, including casting, grinding and polishing the mirrors. His ultimate achievement (1845) was the 'Leviathan of Parsonstown' (this being the old name for Birr). It had a 72-inch mirror, weighing nearly four tons. It had a focal length of 54 ft and a maximum magnification of 6000x. With it he made many important observations 1848–75, especially of nebulae. In these, he discovered many unsuspected details and was able to resolve some of them into stars. He was President of the Royal Society 1848–54. His fourth son, Charles Parsons (1854–1931), also born at Birr, is remembered as the inventor of the steam turbine. (See NEWCASTLE UPON TYNE.)

BISHOP AUCKLAND, Dur.

The Roman fort at Binchester (Vinovia) protected the road between HADRIAN'S WALL and YORK. It dates from AD 80 and was sporadically occupied even after the Roman withdrawal. It covers about four acres of which only a part has been excavated. The house of the Commander contains an elaborate bath suite and a well-preserved hypocaust floor.

BLACKBURN, Lancs.

Home of James Hargreaves (1720/1–1778), a handloom weaver at Stanhill. In 1764, he invented the spinning jenny, which could spin eight threads at once. However, local opposition and the threat of violence led him to move to NOTTINGHAM, where he set up a cotton mill. The Lewis Museum of Textile Machinery (Exchange St.) has an important collection depicting the evolution of textile machinery from the 18th century.

Witton Country Park (Preston Old Road) has a considerable collection of items of agricultural interest. It includes machinery, carriages, wagons and hand tools.

BLACK NOTLEY, Esx. See BRAINTREE

BLAENAU FFESTINIOG, Gynd

The Llechwedd Slate Caverns, and the associated museum, give visitors a dramatic picture of how slate mining was carried out in north Wales. The mine is still operational but now only by open-cast working. Some part of the underground galleries – extending to some 130 miles in all, punctuated by hundreds of huge caverns – have been opened up for visitors. Two tours are available: one in an electrically propelled train and one on foot.

Close by is the Gloddfa Ganol Slate Mine, another major tourist attraction. The old mine had 40 miles of underground tunnels and some of these have been opened to visitors. Associated exhibits include a fine collection of narrow-gauge locomotives.

In 1836 a tramway was built to convey slate to the port of Porthmadog, a dozen miles away, and in 1865 this was converted to a narrow-gauge railway using steam locomotives designed by Robert Fairlie (1831–85). This was closed with the decline of the slate industry but was reopened by a local preservation society and passengers can now travel its full length. Plans are now in hand to built a narrow-gauge link between Porthmadog and Caernarvon following the line of the Welsh Highland Railway, closed 50 years ago.

BLAENAVON, Gwent

There are substantial remains of the old iron works, including five blast furnaces. They commenced operation in the late 18th century and at one time employed 350 people. Two additional furnaces were built in 1810 to meet the demands of the Napoleonic Wars. Production ceased in 1904.

It was at Blaenavon that Sidney Gilchrist Thomas (1850–85) and his cousin Percy Carlyle Gilchrist (1851–1935) developed, in the 1870s, their revolutionary new process for smelting phosphoric iron ore. Among other advantages, this made it possible to make steel from the vast Lorraine/Luxemburg ore deposits. There is an obelisk in memory of Thomas Gilchrist at the site.

The Big Pit Mining Museum displays many aspects of mining, including underground galleries, in which visitors get a very clear view of the old pillar-and-stall working system.

BLAGDON, Som.

Blagdon Reservoir, built in the 1890s, is an important water resource for BRISTOL. The associated Pumping Station originally contained four beam engines, two of which have been preserved. One of them is occasionally operational, but now powered by electricity, not steam.

BLANDFORD CAMP, Dors.

The Royal Signals Museum is devoted to the history of military communications in the British Army from the Crimean War onwards.

BLANDFORD FORUM, Dors.

At Child Okeford, (7 miles NW) is Hambledon Hill, one of the most impressive hillforts in Britain.

BLETCHLEY, Bucks.

Through most of its history Bletchley has been significant only as a major railway junction. During the Second World War, however, it became of immense strategic importance. There was operated, in great secrecy, one of the earliest electronic computers, Colossus, which from 1943 systematically decoded the Enigma-enciphered radio transmissions of the German armed forces. It was well named, for it used 1500 thermionic valves – the invention of the transistor was still years ahead. Because of its importance, moves have been made to preserve the wartime sheds in which it was housed.

BOAT OF GARTEN, Hghld

Until the Beeching closures of 1965, railway travellers in Scotland were familiar with Boat of Garten as an important railway junction. In 1978 the line to Aviemore was reopened as a tourist attraction, with steam locomotives. The Strathspey Railway Museum has been established in the old Highland Railway station. It contains much material of general railway interest, but especially items relating to the old Highland and Great North of Scotland lines.

BODMIN, Corn.

The Trippet Stones (6 miles NE) – probably early Bronze Age – still form an attractive stone circle despite the activities of stone-breakers over the years.

Richard Lower FRS (1631–91), anatomist and pioneer of blood transfusion, was born nearby at Tremeer.

BOLTON, GM

The Hole-i'-the Wood was the home of Samuel Crompton (1753–1827), inventor in 1779 of the spinning mule, so called because it incorporated features of Hargreave's spinning jenny and Arkwright's water frame. The machine revolutionized the fine cotton industry and made Bolton its world centre. Sadly, Crompton's business acumen did not match his inventive genius and he made little profit from his invention. He was born at 10 Firwood Fold, the last remaining thatched cottage in the town.

The Bolton Steam Museum displays stationary steam engines, mostly from the textile industry. The Tonge Moor Textile Museum has a collection of early textile machinery, including Arkwright's water frame, Crompton's mule, and Hargreave's jenny.

Birthplace of J.R. Partington (1886–1965), historian of chemistry. His magisterial four-volume *History of Chemistry* (1961–4) is still the standard work throughout the world. Also the birthplace of Robert Whitehead (1823–1905), inventor of the torpedo.

BO'NESS, Loth.

In the past Bo'ness, on the south side of the Firth of Forth, was an important seaport and centre of a coal-mining area, but both these interests have declined. The Bo'ness Heritage Trust (Bo'ness Station) is committed to recalling the area's history. Its sites include Bo'ness Harbour, Kinneil Colliery and Birkhill Clay Mine, an important centre for the fireclay industry. Hopefully, these industrial museum sites will be linked by a steam railway system operated by the Scottish Railway Preservation Society: steam excursions already operate from the old terminus station.

Kinneil House, once the family seat of the Dukes of Hamilton, was once the home of John Roebuck (1718–94), inventor of the lead chamber process for sulphuric acid manufacture. He was also interested in salt-making, coal mining, and iron smelting, establishing Scotland's first iron works at Carron, later famous for the manufacture of ordnance (carronade). He was elected FRS in 1764.

BONNYBRIDGE, Cent.

Rough Castle (1 mile E) is the remains of a Roman fort on the ANTONINE WALL. One side of the fort is formed by the wall itself; the other three are marked by grass ramparts.

BORDON, Hants.

Headley Mill, still operational, has an exceptionally large (4-acre) mill-pond, which feeds a breast-shot wheel. It dates from the 12th century, but has undergone many structural changes.

BOROUGHBRIDGE, N Yorks.

Famous for the Devil's Arrows: three massive blocks of millstone grit, apparently quarried at Knaresborough, 10 miles away.

BOTALLACK, Corn.

The mining areas around Redruth and Camborne are so littered with the ruins of old engine-houses that they seem just a normal part of the landscape. Two at Botallack, however, are of particular interest, being located on ledges on the cliffs at the extreme tip of Cornwall: the engineering problems of building them must have been formidable.

BOTLEY, Hants.

The Hampshire Farm Museum, based on Manor Farm, depicts the history of agriculture in the county 1850–1950. Apart from displays of agricultural machinery and tools, the exhibits include a reconstructed wheelwright's shop, a forge and a staddle barn.

BOURN, Cambs.

Bourn Mill was built in the 17th century. It has since been somewhat modified but nevertheless demonstrates all the main features of a medieval post mill.

BOURNEMOUTH, Dors.

The Transport Museum displays a collection of trams, buses and trolleybuses.

BOURTON-ON-THE-WATER, Glos.

The Cotswold Motor Museum displays a range of motor-cycles and cars up to about 1950, together with many motoring memorabilia. Also displayed is a reconstructed motor-bicycle workshop of the 1920s.

The Village Life Exhibition is housed in an 18th-century water-mill which was working until 1949. Exhibits include a reconstruction of a blacksmith's forge, a collection of sewing machines and a model of the mill as it was in its working days.

BRACKNELL, Berks.

Home of the Meteorological Office, the national meteorological service. It has a small museum (open to the public by appointment) containing nearly a hundred old scientific instruments. One or two may have been part of George III's collection (now in the Science Museum, London) and others from Kew Observatory.

BRADFORD, W Yorks.

Known as 'Worstedopolis' in the 19th century because of its international standing in the woollen industry. Many mills survive, notably the huge Manningham Mill built by Listers in 1873. At Saltaire (4 miles N) is the model village established in 1851 by Sir Titus Salt (1803–76) to provide decent living and working accommodation for his employees; it included a hospital, school and church. Bradford's museum of the local textile industry – the Industrial Museum, Eccleshill – is located in the old Moorside Mill.

John Mercer (1791–1866) spent most of his working life at the calico-printing firm of Oakenshaw. Without formal education – he began work at the age of nine as a bobbin-minder – he rose to be the acknowledged 'Father of Textile Chemistry' and was elected FRS in 1850. His many inventions included mercerized cotton, in which lustre is conferred on the fibre by treatment with caustic soda.

The National Museum of Photography, Film and Television is the largest of its kind in Britain, devoted to every aspect of image reproduction. The IMAX cinema contains Europe's largest cinema screen.

The Colour Museum (Grattan Road) is located in an old wool warehouse, and, as its name implies, it is devoted to every aspect of colour. The importance of colour in everyday life is illustrated by such themes as colour-blindness, the mixing of colours and the spectrum. A separate gallery is devoted to the role of colour in industry, particularly the textile industry.

The West Yorkshire Transport Museum has a large collection of passenger and goods vehicles, many of them roadworthy and licensed for use.

Birthplace of Sir Edward Appleton FRS (1892–1965), famous for his research on the upper atmosphere and his discovery in 1924 of the Appleton Layer, which reflects radio waves back to earth rather than dispersing outward into space.

BRADFORD-ON-AVON, Wilts.

Centre of an area famous for the production of woollen goods, especially Wiltshire broadcloth. The Industrial Revolution led to the building of large mills, many of which still stand – though converted to other uses – on the waterfront. Typically, the architecture is often extravagant: Abbey Mill (1874) is an example of Venetian Gothic.

Birthplace of Henry Shrapnel (1761–1842), inventor of spherical case shot, first used with effect in the Peninsular War and in the defence of La Haye Sainte on the field of Waterloo.

BRADFORD-ON-TONE, Som.

Cider-making has long been an important industry in Somerset and the Sheppy family have been involved in it since the early 19th century. The Sheppy Museum features old cider-making equipment and the tools of associated craftsmen, notably coopers.

BRADING, IoW

The Roman villa here was excavated at the end of the 19th century and several fine 3rd-century mosaic floors are displayed. A small associated museum displays a variety of artefacts unearthed in the course of the excavation.

BRADWELL, Esx

Nuclear Electric's massive power station dominates the local landscape. It incorporates two Magnox reactors with an output of 245 MW.

BRAINTREE, Esx

The town has long been a centre of the silk industry and Samuel Courtauld (1798–1881) built a factory there in 1817. The main product was black crape, a coarse silk much in demand by Victorian widows during their prolonged periods of mourning. In 1904, they acquired the patent rights for viscose rayon ('artificial silk') and began to manufacture it in COVENTRY.

John Ray (1628–1705), the natural historian, was born at Black Notley (3 miles S) where his father was a blacksmith. With Francis Willughby (1635–72) he began a systematic description of the entire plant and animal world, but Willughby died before it was completed. Ray went on to produce his famous *Historia Generalis Plantarum* and many other important botanical and zoological works. He is buried in the churchyard of SS Peter and Paul, where a plaque was put up by the Ray Society in

1984. There is also a monument to Ray in Braintree where he attended the grammar school before going to Cambridge. He spent the last years of his life at Black Notley.

BRAMBER, W Ssx

The history of the tobacco industry – growing, processing and marketing – is well-recorded. The House of Pipes in Bramber – a collection of nearly forty thousand items from all over the world – is devoted to the use of tobacco. Exhibits include pipes, snuff-boxes, tobacco jars, cigarette holders, lighters and countless other memorabilia.

BRAUNTON, Devon

Few historic sites can be recommended as having nothing dramatic to show. This paradox prevails at Braunton, one of the very few places where the medieval system of open-field cultivation still prevails.

BREAMORE, Hants.

The Breamore Countryside Museum, in the grounds of Breamore House (1583) – home of the Hulse family for ten generations – depicts the history of village life from the early 19th century. Exhibits include a reconstruction of a forge, a brewery, a cider factory, saddler's shop and a shoemaker's workshop. A number of horse-drawn vehicles are displayed, including the Red Rover stagecoach which served the Southampton–London run.

BRECHIN, Tays.

Birthplace of Sir Robert Watson-Watt (1892–1973), pioneer of radar, the practicality of which he first demonstrated in 1935. When war broke out in 1939, Britain had a virtually complete national network of radar stations to indicate the approach of hostile aircraft, and this was of crucial importance in the Battle of Britain in 1940, a turning-point in the Second World War.

BRENTFORD, GL

In 1975 British Leyland established the British Motor Industry Heritage Trust Museum to bring together the large collection of vehicles it had acquired over the years from its various constituent companies. About a hundred of the 300 vehicles are on display at any given time. Makes range from Albion to Wolseley, Jaguar to Riley.

The Musical Museum (High St) displays a variety of mechanical musical instruments. These include a Mighty Wurlitzer from Chicago, a barrel-organ and musical boxes.

BRESSINGHAM, Nflk

The Bressingham Live Steam Museum displays standard-gauge locomotives, stationary engines, a steam roundabout, fire-fighting equipment and hundreds of exhibits related to the history of steam.

BRIDGNORTH, Shrops.

Once an important port on the River Severn and famous for carpet manufacture; many waterside warehouses still survive. The Severn Valley Railway (Shrewsbury to

Hartlebury) was closed in the 1960s but the section from Bridgnorth to Kidderminster has been reopened by a preservation group and has an exceptional collection of rolling stock. A fine cast-iron bridge designed by John Fowler (1817–98) – who also designed the Forth Bridge – takes the railway over the Severn at Eardington. A cliff railway built in 1892 links the Low Town and the High Town.

The Midland Motor Museum (on Stourbridge Road) specializes in sports and racing motor-cycles and cars, mostly post–1920. The Northgate Museum (High Street) is largely devoted to local history but there are some exhibits of wider interest. These include an 18th-century turret clock and a model of a steam-pumping engine made to the design of Richard Trevithick (1771–1833).

BRIDGWATER, Som.

Home of John Allen (1670–1741), physician. His *Synopsis Universae Medicae* (1719) ran to many editions and was widely used in Britain and on the Continent in the first half of the 18th century. He was also a versatile inventor, devising an engine 'to raise water by fire' and a steel-sprung road carriage. He published details of them in his *Specimina Iconographia*, a copy of which he presented to George II.

BRIDPORT, Dors.

At Askerswell (5 miles W) is Eggardon Camp, one of the most impressive of British hillforts.

BRIERFIELD, Lancs.

The Neville Blakey Museum of Locks, Keys and Safes displays a remarkable historical collection of items in this field.

BRIGHTON, E Ssx

The Regency Pavilion (1784) is primarily of architectural interest but is of some significance in relation to the gas industry. In 1821, the Prince Regent had gas installed for lighting the Music Room and the Banqueting Hall. This royal patronage was a valuable stimulus for the new industry – The Gas Light and Coke Company had been formed less than ten years previously.

The first railway using electric traction was demonstrated at the Berlin Exhibition in 1879. In 1883, Magnus Volk opened the first electric railway in Britain, on the front at Brighton.

Birthplace of Sir Martin Ryle (1918–84), pioneer of radio astronomy. He was Astronomer Royal 1972–82 and was awarded a Nobel Prize in 1974.

BRILL, Bucks.

The villagers are reputedly taciturn but the restored 17th-century post mill immediately catches the eye without need for direction. According to local doggerel:

> I went to Noke and nobody spoke,
> I went to Brill it was silent and still,
> I went to Thame it was just the same,
> But I went to Beckley and they answered directly.

BRISTOL, Avon

An important West Country port, industrial centre and university city. Its first claim to scientific fame was in the late 15th century when Thomas Norton – a member of a well-known local family and elected MP for the borough in 1436 – took up alchemy and about 1477 wrote his *Ordinall of Alchimy*. It was published in Latin in 1618 and in English in 1652. The author was supposedly anonymous but a simple cipher reveals him as:

> Thomas Norton of Bristo
> A parfet master ye maie call him trowe.

Like his contemporary philosophers, his aim was the elixir of life and the transmutation of base metals into gold, but the book contains much practical information on chemical manipulations.

A near-contemporary of Norton was Sebastian Cabot (1474–1557) who was born in Bristol and returned there in 1547 after a tempestuous career as a merchant venturer. The prominent 105-ft Cabot Tower on Brandon Hill was built to commemorate the fourth centenary of his discovery of the North American continent in 1497/8. A replica of his ship, the *Matthew*, has been built at Redcliffe Wharf to celebrate the 500th anniversary of this epic voyage. He was a skilled navigator and map-maker and for a time was employed as such in Seville. In 1544, he published an engraved map of the world (now lost) and he investigated the variation of the compass needle. A statue of Cabot stands outside the Arnolfini Gallery on Narrow Quay. The geographer Richard Hakluyt (1522–1616) was for a time prebendary of Bristol.

In 1798, Thomas Beddoes (1760–1808) founded in Clifton, a suburb of Bristol, a Medical Pneumatic Institute (plaque in Dowry Square) to promote the medical use of gases, or 'airs' as they were then called. After a rather stormy career, he had been appointed Reader in Chemistry in OXFORD in 1788: he claimed to have attracted the largest classes in the University since the 13th century. However, in the Long Vacation of 1792 he unwisely distributed a pamphlet sympathetic to the French Revolution and was obliged to leave Oxford. He went to practise medicine in fashionable Clifton and it was from this that the idea of the Institute arose. In the event it was not a success and about 1801 he closed it and started afresh in London. Nevertheless it did have very important consequences. In 1798, he appointed a young Cornishman, Humphry Davy (1778–1829) – apprenticed to a surgeon in PENZANCE – as superintendent. Beddoes not only encouraged Davy's incipient interest in chemistry but introduced him to his wide circle of literary friends, including James Watt and the poets Southey, Wordsworth and Coleridge. He thus acquired social graces which stood him in good stead when, after little more than a year, he joined the Royal Institution in LONDON, a very fashionable scientific centre, as Professor of Chemistry; this was the start of a brilliant scientific and social career. A memorial plaque on Beddoes' house in Rodney Place commemorates both him and Davy.

Another chemist with Bristolian connotations was William Prout (1785–1850) born near Chipping Sodbury. He is remembered for Prout's Hypothesis (1815) formulated in the aftermath of John Dalton's revolutionary theory (1803) that the atoms of different elements are distinguished by having different weights. According to Prout all atomic weights were simple multiples of that of hydrogen, the lightest element known. This stimulated much precise analytical work, which seemed to prove him wrong. However, he was to some extent vindicated a century later with the discovery of isotopes, which do in fact have approximately whole-number atomic weights.

In more recent times, the centre of scientific activity in Bristol has been the University, founded as a university college in 1876 and accorded full university status in 1909. Here William Ramsay (1852–1916) was appointed Professor of Chemistry in 1880; seven years later he moved to University College, London, where he continued the brilliant research that led to the discovery of a whole new family of chemical elements, the inert gases of the atmosphere. For this he was awarded a Nobel Prize in 1904. His principal collaborator was M.W. Travers (1872–1961), who was appointed to the same professorship in Bristol as Ramsay had once held.

One of the great industries in Bristol is tobacco and the Wills family have long been generous benefactors. One of their memorials is the H.H. Wills Physics Laboratory which has a reputation in particle physics ranking with the Cavendish Laboratory in CAMBRIDGE. Here C.F. Powell (1903–69) did pioneer work in recording atomic particles by the tracks they leave in photographic film. In 1947 this led to the discovery of a new particle, the pion. One of his colleagues was N.F. Mott, who was Director of the Laboratory from 1948 to 1954, when he was appointed to the prestigious Cavendish Professorship of Experimental Physics at CAMBRIDGE. He was awarded a Nobel Prize in 1977.

Nevertheless, pride of place among Bristol-born scientists must be accorded Paul Dirac (1902–84), son of a Swiss-born teacher. Dirac, who ranks with giants such as Newton and Einstein, was a theoretical physicist and a founder of quantum mechanics, who predicted the existence of the positron and other antiparticles. He graduated in electrical engineering at Bristol before going on to Cambridge where he became Lucasian Professor of Mathematics, a position once held by Newton. He was awarded a Nobel Prize in 1933. A memorial to him was unveiled in Westminster Abbey in 1996.

Among Bristolians famous in the medical sciences were the physicians Richard Bright (1789–1858), discoverer of Bright's disease and William Budd (1811–80), pioneer epidemiologist who practised there 1842–73. Budd was physician to Bristol Royal Infirmary and in 1873 showed that typhoid is spread by contagion. There is a memorial plaque on his house, 89 Park Street (now George's Bookshop). More colourful was Thomas Dover (1660–1742) who started to practise there in 1682. In 1708 he set out on a privateering voyage under Captain Woodes Rogers in the course of which the castaway Alexander Selkirk (Defoe's Robinson Crusoe) was discovered. He is famous for Dover's Powder, a fearful nostrum compounded of equal parts of opium, ipecacuanha, liquorice, saltpetre and cream of tartar. There is a memorial plaque in Dover Place, named after him. In the same context, two other Bristolians deserve mention. Elizabeth Blackwell (1821–1910) was the first woman to be placed on the British Medical Register (1859): there is a memorial plaque at 1 Wilson Street. A.V. Hill (1886–1977) was noted for his research on the physiology of nerves and muscles, and won a Nobel Prize in 1923.

Bristol is a notable centre for technology as well as science. Ceramics have long been important and in 1770 William Cookworthy (1705–80) established his pottery on Castle Green to make some of the earliest European ware in the style of Chinese porcelain. The old blue glass of Bristol is prized by collectors world-wide and is once again being made at the Glass Works in Colston Yard. In 1831, Peregrine Phillips, a local vinegar manufacturer, patented a novel process for making sulphuric acid, arguably the most important of all industrial chemicals. Sadly, it was ahead of its time and Phillips disappeared into obscurity, but half a century later it was reinvented and became one of the most important of all industrial chemical processes. In quite a different field was William Friese-Greene (1855–1921), born at 69 College Street. As a young man he became interested in photography and in 1875 he moved to BATH to open a studio, and then on to LONDON. There he made a name for himself as a pioneer of cinematography. There is a memorial to him at the Orpheus Cinema.

Bristol has very close connections with Isambard Kingdom Brunel (1806–59), the great engineer. He designed the famous Clifton Suspension Bridge which dramatically spans the Avon Gorge at Clifton – although it was not completed until 1864, after his death – and laid out the route for the Great Western Railway from Paddington: the magnificent Temple Meads Station is a lasting memorial to him. Appropriately, Brunel's original GWR terminus has been restored and turned into a hands-on science centre, the Exploratory. He also built, in 1843, the SS *Great Britain*, an iron-hulled, screw-propelled vessel which in its day was the largest ship afloat. She was launched by Prince Albert, who arrived at Temple Meads from London in the then record time of 3 hours 40 minutes. She ended her days as a hulk in the Falkland Islands but was eventually brought back to Bristol to the very dock in which she had been built, where she is being painstakingly restored. Another great engineer associated with Bristol was the road engineer John Loudon McAdam (1756–1836), Surveyor-General to the Bristol Roads Trust 1815–27, and inventor of the 'macadamizing' system of road construction. (See AYR.)

Close by the SS *Great Britain* is the Maritime Heritage Centre, devoted to the history of shipbuilding in Bristol. In this, paddle-steamers were important in the 19th century: because of their shallow draught they could navigate the Avon both ways on the same tide. The splendidly restored *Waverley* and *Balmoral* – the last sea-going paddle-steamers in the world – regularly ply the Severn Estuary on pleasure cruises.

The Underfall Yard marks the site of the original embankment which made possible the Floating Harbour, where ships could remain afloat whatever the state of the tide. In 1831 a system of culverts was designed by Brunel for removing mud from the harbour. Exhibits at the Maritime Heritage Centre include a full-size replica of Brunel's scraper-dredger (1832). (See LONDON.)

Still in the world of shipping, Bristol was the birthplace of Samuel Plimsoll (1824–98), the 'Seaman's Friend'. As an MP he was instrumental in effecting the Merchant Shipping Act of 1876. One provision was that all merchant ships should carry the Plimsoll Mark to indicate their loading limit. There is a bust of Plimsoll at Cumberland Basin, Portway, and a plaque at 9 Colston Parade. (See LONDON.)

A very different kind of industry has long been important in Bristol. In 1748 John Fry (1728–87) began to manufacture chocolate in the centre of the city: like the Cadburys of Bournville (see BIRMINGHAM) and the Rowntrees of YORK, the Frys were a Quaker dynasty, but marginally the least successful of the three. After the First World War – ruinous to the whole chocolate industry, with imports restricted and exports impossible – Frys merged with Cadbury to form the British Cocoa and Chocolate Co., though both continued to trade separately. In 1921, the business was moved out of the city centre, which it had long dominated, to a new factory at Somerdale, halfway between Bristol and BATH.

The huge bonded warehouses dominating the landscape at Ashton Gate are a reminder of Bristol's historic association with the tobacco industry. Once snuff was an important product and many snuff-mills were built on high ground within and near the city. One such was located on Clifton Down near the Suspension Bridge. It was destroyed by fire in the 1820s and rebuilt in 1829 as an observatory: it now houses a camera obscura providing panoramic views of the local scene. A cave below opens on to the cliff face of the Avon Gorge, giving a fine view of the bridge.

The most recent of major industries in Bristol is aeroplane and aero-engine design. The Bristol Aeroplane Company (now part of British Aerospace) was established at Filton in 1909, the year in which Louis Blériot dramatically demonstrated the potential of air transport by making the first flight across the English Channel. Many famous aircraft have been developed here, ranging from the Bristol Fighter of the First World War to Blenheims and Beauforts in the Second, and perhaps now the best known of all, the Anglo-French Concorde. Bristol-built aero-engines and aircraft from 1910 onwards are displayed at the Bristol Industrial Museum, Canons Marsh, as part of a much wider display of the city's industrial and transport history.

Botanic gardens – gardens laid out not to please the eye but for the systematic study of plants – date back to classical times, but particularly to the rise of herbalism in the 16th century. In the university context, they are important for providing material for teaching and research. Bristol University's botanic garden, formerly at Tyndall Avenue, is now located at Bracken Hill, beyond the Suspension Bridge. Bristol Zoo (Clifton), like virtually all zoos today, is primarily a visitors' attraction but the collection is extensive and well-organized and thus useful also to the serious student.

BRIXHAM, Devon

The British Fisheries Museum is an outstation of the National Maritime Museum, Greenwich, and depicts five centuries of deep-sea fishing. Fishing and shipbuilding are also depicted in the Brixham Museum (Bolton Cross).

BROADSTAIRS, Kent

Birthplace of the engineer Thomas Russell Crampton (1816–88), responsible for laying the first Dover–Calais submarine telegraph cable (1851) and remembered as designer of the innovative 'Crampton' railway locomotive. (See LIVERPOOL.)

BROADSTONE, Dors.

Old Orchard House was the home for the last eleven years of his life of Alfred Russel Wallace (1823–1913), distinguished natural historian, who shared with Charles Darwin (1809–82) the formulation of the theory of biological evolution in a joint paper presented to the Linnaean Society in London in 1858. His grave is in Broadstone Cemetery. (See USK.)

BROADWAY, Glos.

Snowshill Manor (National Trust) lies about 3 miles SW on the B4632. Its large random collection includes agricultural implements, material relating to weaving and lace-making, and clocks, watches and some scientific instruments.

BROMHAM, Beds.

Bromham Water Mill, now restored to working order, houses a small museum of milling history.

BROMSGROVE, H & W

The Avoncroft Museum of Buildings displays various industrial buildings re-erected on the site, including a chain works and a nailer's workshop. There is also a well-preserved post mill.

BROUGHTON, Cumb.

Swinside (2 miles NW) is a large and nearly perfect stone circle; thirty-two stones are still upright out of a total of fifty.

The blast furnace at Duddon Bridge (2 miles W), built in 1736, is remarkably well preserved.

BROUGHTY FERRY, Tays.

Broughty Castle Museum is devoted to local history and as such reflects the industry of the area, including fishing. A section on the whaling industry, of which nearby DUNDEE was an important centre, is of particular interest.

BRUREE, Lim.

Eamon de Valera is best known as a politician and first President of Ireland, but he was also a competent mathematician with a lifelong interest in the subject. He did much to promote mathematics and science in DUBLIN. His home, De Valera Cottage (Knockmore) has been restored to its original state and the associated museum displays the largest collection of memorabilia in Ireland.

BUCKHAVEN, Fife

The history of the local fishing industry, and the equipment used, are depicted in the Buckhaven Museum.

BUCKIE, Gramp.

The Buckie Maritime Museum has exhibits depicting the history of the local fishing industry, and also coopering.

BUCKINGHAM, Bucks.

The Buckingham Movie Museum (Market Hill) is unusual in being devoted not to professional film-making but to filming by amateurs. It possesses over three hundred cameras and projectors, with many associated memorabilia.

BUDE, Corn.

The Bude Canal, designed for tub-boats, was built in the early 19th century to link the

River Tamar with the English Channel, with access – via a sea-lock – to the Bristol Channel. The Bude Historical and Folk Exhibition, housed in part of the old canal buildings, depicts the history and working of the canal.

BUDLEIGH SALTERTON, Devon

There has been a corn-mill on the River Otter, at Otterton (3 miles N), since the 10th century. The present mill is the result of rebuilding in the 1840s. It is still a working mill and includes a small museum devoted to the use of water-power in milling.

A complementary display is devoted to lace-making, a cottage industry particularly identified with HONITON.

BUNRATTY, Clare

The present Bunratty Castle dates from 1467 and was restored in the 1950s by Lord Gort, who presented it to a managing trust. The adjoining Folk Park includes replicas of workshops from various parts of Ireland: the thatcher's craft is well-represented.

BURFORD, Oxon.

The Cotswold Wildlife Park has a large well-documented collection of animals – ranging from reptiles and birds to large cats and rhinos – on view in two hundred acres of parkland.

BURNLEY, Lancs.

The mile-long embankment carrying the Leeds and Liverpool Canal across the Calder Valley was completed in 1799.

The Weavers' Triangle Visitor Centre (Manchester Road) is situated in an old toll-house on the Leeds and Liverpool Canal. It includes examples of all the buildings – foundries, weaving sheds, warehouses, and so on – used in the heyday of Burnley as a cotton town in the 19th century.

Queen Street Mill (Harle Dyke) dates from 1894. It is an operational steam-powered cotton mill maintained by Pennine Heritage.

BURREN, Clare

The Burren uplands are rich in megalithic remains. Two of the most outstanding are the wedge tomb Poulaphuca (cave of the spirits) and the massive Poulnabrone dolmen. The latter consists of a huge tilted capstone, supported by four uprights.

BURROWBRIDGE, Som.

Burrowbridge Pumping Station was built for land drainage in 1869. Two of the original Easton steam engines, in use until 1955, survive and two others (1864 and 1869) have been installed from other sites.

BURSTOW, Sry

John Flamsteed (1646–1719), first Astronomer Royal, was rector here for more than thirty years, but delegated most of his duties. There is a memorial to him in the chancel.

BURTON UPON TRENT, Staffs.

The Bass Museum of Brewing illustrates the history of brewing and transportation of beer, and related equipment such as pumping engines. The Heritage Brewery Museum, located in 19th-century buildings, is a working museum of brewing.

BURWASH, E Ssx

The Bateman Estate is particularly remembered as the home of Rudyard Kipling (1865–1936), who settled there in 1902. Bateman's Mill, now restored and used to grind flour, dates from the 18th century. Close by is a water turbine installed by Kipling in 1903 to generate electricity.

BURY, GM

Home town of John Kay (1704–64) inventor in 1733 of the flying shuttle which made it possible for one weaver to do the work of two. In 1753, the Bury Mob destroyed his works and he retired to France, where he died in poverty.

The Transport Museum (Castlecroft Road) displays locomotives and rolling stock, together with much railway equipment.

BURY ST EDMUNDS, Sflk

Angel Corner, a Queen Anne House (National Trust) houses the Gershon-Parkington Memorial Collection of Clocks and Watches, dating from the 16th century.

BUSHMILLS, Ant.

The Giant's Causeway is a world-famous geological formation and tourist attraction. It has been designated a World Heritage Site by UNESCO. It consists of a mass of some forty thousand stepped basalt columns, mostly hexagonal; some are 40 ft high. They seem first to have been studied by the French geologist Nicolas Desmaret (1725–1815) who recognized their similarity to outcrops in the Auvergne. The English geologist John Whitehurst (1713–88) identified them as igneous intrusions of volcanic origin.

BUXTON, Derbs.

Birthplace of James Brindley (1716–72), who, despite being no more than semi-literate, became a great canal builder, sometimes in association with the Duke of Bridgewater.

BYERS GREEN, Dur.

Home of Thomas Wright (1711–86), a competent astronomer who sought to interpret the results of observation with God's place in the Universe. He embodied his ideas in *An Original Theory or New Hypothesis of the Universe* (1750) and correctly concluded that the Milky Way must be disc-shaped. He built himself an observatory in nearby Westerton; this still survives as Westerton Folly.

C

CADBURY, Devon

Strong hill fort (on A3072 Crediton–Tiverton road) with ramparts in places 6 m high.

CAERLEON, Gwent

In its day the Roman fort at Caerleon (Isea) had a garrison of five thousand men. Much of it has been excavated, revealing a fine amphitheatre, barracks and baths. The National Museum of Wales (CARDIFF) has mounted a good interpretative exhibition.

CAERNARFON, Gynd

The castle, built 1283–1330, is unusually well-preserved. Earlier the Romans had a fort at Segontium, overlooking the town. The Segontium Roman Fort Museum has exhibits interpreting the conquest and occupation of Wales by the Romans.

The Seiont II Maritime Museum records the history of Caernarfon as a port from Roman times, with special reference to the slate industry. Its most striking exhibit is the *Seiont II* steam dredger, fully restored, and the *Nantlys* (1920), the last of the many ferries that plied across the Menai Strait from the 17th century: she was laid up in 1976.

CAERPHILLY, Gwent

While military technology is well-represented in museums by hand weapons, armour and firearms, siege weapons – widely used since classical times – are less frequently exhibited. Caerphilly Castle displays four full-scale (replica) siege engines. The castle (built *c*. 1268) covers 30 acres and in the whole of Britain is matched only by Windsor and Dover.

CAISTER-ON-SEA, Nflk

Caister Castle was the home of Sir John Fastolf (1378–1459), whose family lived here for several generations: he is supposedly the original of Shakespeare's Falstaff. It now displays a variety of steam and petrol-driven road vehicles, including a Panhard-Levassor of 1893.

CALBOURNE, IoW

There was a water-mill at Calbourne from the 11th century until 1955, and the 20-ft overshot water-wheel is still operational. A supplementary steam engine was installed in 1894 to power a roller-mill. Apart from the mill and its machinery, the Calbourne Water Mill and Rural Museum displays a variety of ancillary equipment, including grinders for maize and beans, dairy equipment and a wheat cleaner.

CALNE, Wilts.

Bowood House (3 miles W) is the country home of the Shelburne family. During the period 1773–80, the chemist Joseph Priestley (1773–1804) was employed by the Earl of Shelburne as librarian. Here in 1774, in a room set aside for him as a laboratory, and still to be seen, he discovered oxygen.

CAMBERLEY, Sry

Birthplace of F.W. Twort FRS (1877–1950), bacteriologist remembered for his discovery (1915) of bacteriophages, viruses which in order to reproduce infect and destroy bacteria.

CAMBORNE (now combined with REDRUTH), Corn.

A great centre of the old Cornish tin-mining industry. At Pool (2 miles W), there is a big collection of beam engines used for pumping water from depths up to 600 m. The Poldark Mine and Heritage Complex (on the Redruth–Helston Road), offers underground tours and museums of Cornish mining. The Camborne School of Mines, Redruth, has a large collection of minerals and ores.

Richard Trevithick (1771–1833), pioneer of steam traction, was born nearby at Illogan – where his father was manager of Dolcoath Mine – and went to school in Camborne. The cottage where he lived still stands. At a house in Redruth, William Murdock (1754–1839) in 1792 installed one of the first systems of gas lighting.

Tuckingmill was the home of William Bickford (1774–1834) inventor of the safety fuse for explosives (1831). Disturbed by the hazards of the shot-firing methods then in use, he devised a fuse in which a light cord was spun round a central core of gunpowder. He set up a factory to manufacture the fuse; this closed in 1961. He is buried in the graveyard of Tuckingmill Church.

At Pool can be seen an impressive beam engine that demonstrates the enormous size these ponderous engines eventually reached. Built by Harveys of Hayle in 1892, it was moved to its present site in the 1920s. The beam alone weighed 50 tons and the cylinder was 9 ft in diameter. Although designed to work at only five strokes a minute it could raise 24,000 gallons of water an hour from a depth of 1500 ft.

CAMBRIDGE, Cambs.

Although some peripheral light industry has developed in recent years, Cambridge is still pre-eminently a university city, although the autocratic powers of the University have been greatly diminished over the last century or so. The Great Eastern Railway, which reached the city in 1845, announced its intention to run excursion trains on Sundays. This potential incursion of 'foreigners and undesirable characters . . . on that sacred day' so incensed the University that the Vice-Chancellor wrote:

> The Vice-Chancellor of the University of Cambridge wishes to point out to the Directors of the Great Eastern Railway that such a proceeding would be as displeasing to Almighty God as it is to the Vice-Chancellor of the University of Cambridge.

So much for the Industrial Revolution! Today, the complaint is that the station was built too far from the city centre.

Consequently, the scientific and technological connotations of Cambridge lie largely within the University and – despite the wealth of literature, both scholarly and touristic – something needs to be said about its evolution. A few religious foundations were already in existence in the 12th century and it seems that it was this that attracted a small group of Oxford scholars in 1209 when town-and-gown rioting had brought their own university virtually to a standstill. The little community prospered and expanded and became recognized as a *studium generale*, an organized place of study to which students from all parts were admitted. In 1284 a development occurred that was to determine its organization for teaching and research for the next six centuries. In that year, the first of the great Cambridge colleges, Peterhouse, was founded and is an interesting example of academic cross-fertilization. It was ordained that scholars should live together 'according to the rule of the Scholars at Oxford who are called Merton'. A much later foundation, Trinity College (1546), was similarly to become the model for Trinity College, DUBLIN. In the meanwhile, largely in response to formal recognition of Cambridge by the Pope as a university (1318), twelve other colleges had been founded, among them Clare, Pembroke, Corpus Christi, Jesus and Christ's. This great era ended with the foundation of Sidney Sussex College in 1596. Not until the 19th century and later were there any new foundations of consequence. Girton, the first college for women, was founded in 1869 but it was nearly a century (1948) before women were admitted to full membership of the University. Churchill College was founded in 1960, by public subscription, as a memorial to Winston Churchill, and was intended to be devoted largely to mathematics, engineering and science.

The foundation of the colleges – for the most part developing into well-endowed and influential institutions – changed the character of the University, which became in effect a federal body with power residing in the colleges: these were individually self-governing within general university statutes. Not until the 20th century did the University as such regain the ascendancy. After the First World War, a Royal Commission on the universities of Oxford and Cambridge was set up and new statutes were approved in 1926. In particular, it was recognized that existing endowments were insufficient to meet existing demands, especially in science: for the first time, substantial financial help came from the state. Over the years, state aid has been manifested in a number of well-equipped laboratories for all the major branches of science and engineering, although in many cases there have also been substantial private subventions, notably from industry and commerce. For around the first five centuries of the University's history, scientific achievement was necessarily identified with individual colleges – because that was where the action was – but today it is in the university laboratories that progress mainly lies.

Although Peterhouse was to be governed on rules similar to those of Merton College, OXFORD, it does not seem that that college's early interest in mechanics, geometry and physics was translated to Cambridge. One of the first scientists, in the modern sense, to be identified with Cambridge was William Gilbert (1544–1603) of COLCHESTER, who graduated in arts and medicine and was for a time a Fellow of St John's. Later he practised medicine in London and became President of the College of Physicians. His real claim to fame, however, is as an experimental investigator of magnetism and electricity, phenomena then little understood. His *De Magnete* (1600) is a classic in its field and was acclaimed by, among others, Galileo and Francis Bacon. Bacon (1561–1626), founder of the Baconian system of inductive reasoning, was a student at Trinity before embarking on a political career. William Harvey (1578–1657) of FOLKESTONE, also studied medicine at Cambridge, at Caius College, and then went on to study at Padua under the great anatomist Fabricius. He then became a successful physician in London and achieved fame for his discovery of the circulation of the blood. His *Anatomical Treatise on the Movement of the Heart and Blood in Animals* (1628) is one of the most significant medical works of all time. He also did highly original research on embryology.

But the real flowering of science in Cambridge really dates from Isaac Barrow (1630–77), who had a long connection with Trinity, latterly as Master: he was finally Vice-Chancellor of the University. He is remembered today for his writings on mathematics and optics and by his contemporaries for his long sermons and slovenly habits. In 1663, he was appointed first Lucasian Professor of Mathematics and in this capacity was tutor to Isaac Newton (1642–1727) whose gifts he soon recognized as greater than his own: in 1669 he resigned his chair in Newton's favour.

Newton had entered Trinity in 1661 and seems not to have been a noticeably brilliant student. His geometry, in particular, was rated poor but nevertheless he went on to be one of the greatest mathematicians of all time. Due to the Plague he spent the years 1665–6 away from Cambridge (see GRANTHAM) but used the time to formulate his ideas on optics, mathematics and dynamics. His *Principia* – made possible by his brilliant solution to the problem of Kepler motion – was published in 1687. In optics his great discovery was that white light is in fact a mixture of seven different colours. His *Optics* (1704) was a masterly exposition of experimental science. Newton's revolutionary ideas quickly dominated European thought. Unfortunately, his genius was marred by a quarrelsome nature, which led him into acrimonious disputes with many of his contemporaries. He became dissatisfied with academic life and in 1696 accepted appointment as Warden of the Mint in London. He left Cambridge finally in 1701 and two years later was elected President of the Royal Society, an appointment he held until his death. Statues of both Barrow and Newton are in the entrance to the chapel at Trinity.

Gonville and Caius College is a memorial to John Caius (1510–73) who in 1557 endowed and enlarged the existing Gonville Hall, where he had been a student. He finally qualified in medicine at Padua, studying anatomy under Vesalius, and then established himself as a lecturer on anatomy and a successful physician in London, where he was nine times elected President of the College of Physicians. A frieze on the front of the college depicts a number of famous members including Caius and Harvey, and W.H. Wollaston (1766–1828), pioneer of platinum metallurgy. Caius' tomb in the college chapel bears the simple memorial *Fui Caius* (I was Caius). Another famous medical graduate of Caius was C.S. Sherrington (1857–1952) justly called 'the Harvey of the nervous system'; his reputation, however, was made at OXFORD. His contemporary at Cambridge was E.D. Adrian (1889–1977) who made major contributions to our understanding of how nervous impulses are conducted. He was Master of Trinity 1951–65 and President of the Royal Society 1950–55. As Baron Adrian of Cambridge he was one of Britain's last hereditary peers. In 1932 he and Sherrington shared the Nobel Prize for Physiology or Medicine. Yet another distinguished member of Caius was the self-taught mathematician George Green (1793–1841). His *Essay on the Application of Mathematical Analysis to the Theories of Electricity and Magnetism* (1828) passed unnoticed at the time but later so profoundly impressed Lord Kelvin at GLASGOW that he had it reprinted after Green's death.

The 19th century saw many distinguished scientists in Cambridge. One was the geologist Adam Sedgwick (1785–1873): he, too, was a Trinity man and mainly taught mathematics. In 1818, however, in the manner of the time, he was appointed Professor of Geology, a subject of which he knew virtually nothing. Nevertheless, the choice proved a good one, for he became one of the most eminent geologists of his day. The Sedgwick Museum in Pembroke Street is his memorial. During part of Sedgwick's professorship, the Master of Trinity was the polymath William Whewell (1794–1866), who was Professor of Mineralogy. On balance, he was more interested in the philosophy of science than in experiment, and his *Philosophy of the Inductive Sciences* (1838) was widely read. It was he who coined the word 'scientist' in its modern sense, as well as many other terms in the scientific vocabulary.

At the same time another scientist of international stature was preparing himself for life – in the Church, as he supposed. This was Charles Darwin (1809–82) who entered Christ's College in 1827 after desultorily studying medicine in Edinburgh. However, natural history proved more attractive than Holy Orders and in consequence he became friendly with the botanist J.S. Henslow (1796–1861) at St John's. Henslow had just been appointed Professor of Botany and was busy revivifying its teaching at Cambridge and restoring the neglected botanic garden. This was to change the whole course of Darwin's life, for it led to his going as unpaid naturalist on the five-year circumnavigation of the world in HMS *Beagle*, during which he formulated the basis of his theory of evolution. There is a bust of Darwin in the college but he is a good example of a prophet being not without honour save in his own country. Although he was eventually accorded the highest honours by foreign governments, his own gave him nothing save burial in Westminster Abbey.

However, the Darwin name lives on in Cambridge. Of his sons, George Darwin (1845–1912) was for 30 years Professor of Astronomy and Francis Darwin (1848–1925) was a botanist. Horace Darwin (1851–1928) followed a different bent, as an engineer: he founded a small scientific instrument manufacturing firm which eventually (1881) became the Cambridge Scientific Instrument Company. This prospered, partly because of the high quality of its products and partly because of the rapidly growing demand for laboratory apparatus of all kinds. Among the company's customers were James Clerk Maxwell, Lord Rayleigh, Francis Galton, and Kew Observatory. One of the staff at Kew was H.S. Whipple (1871–1953) who became Horace Darwin's personal assistant in 1898 and eventually chairman of the company. He was a keen collector of old scientific instruments and books, and in 1944 presented these to the University. They are now displayed in the Whipple Museum in Free School Lane. The family name is perpetuated in Darwin College – founded in 1964 for postgraduate students – built around George Darwin's old home in Silver Street.

By mid-century, astronomers were looking deeper and deeper into space, but the solar system still held many surprises. In 1841, John Couch Adams (1819–92) – then newly arrived as a student of mathematics at St John's – deduced that some detected irregularities in the orbit of Uranus were due to the existence of a still undiscovered planet. By 1845, he had calculated the position of the planet and conveyed the information to Sir George Airy (1801–92), Astronomer Royal. Unluckily the message miscarried. Meanwhile, the French astronomer Urbain Le Verrier (1811–78) had reached the same conclusion and on the basis of his calculations the new planet, Neptune, was observed at the Berlin Observatory. Inevitably this led to bitter controversy. Adams devoted the rest of his life to the dynamics of the solar system, on which he became the leading authority. He was appointed professor of astronomy and geometry in 1858. (See TRURO.)

Newton's discoveries influenced scientific thought throughout the world for some two centuries, and nowhere more so than at Cambridge. George Stokes (1819–1903), one of his successors as Lucasian Professor of Mathematics, spent his whole working life in Pembroke College, where he finally became Master in 1902. He did important work on fluid dynamics: he is remembered by Stokes' Law, governing the rate of fall of solid bodies in viscous liquids. But the real resurgence of mathematical physics came in the 1870s when James Clerk Maxwell (1831–79) was appointed the first Cavendish Professor of Experimental Physics and was charged with designing the original Cavendish Laboratory – now perhaps the best-known laboratory in the world. Maxwell, with his electromagnetic theory of light, can be said to have initiated a revolution in physics – continued by other giants such as Max Planck and Albert Einstein – as profoundly significant as that of Newton in dynamics two centuries earlier but no more than a mile away. For nearly a century the Cavendish Laboratory was housed in its original building in Free School Lane but in 1974 it moved to fine new buildings in Madingley Road.

Sadly, Maxwell died at the height of his powers at the early age of 48. Had he lived to his allotted age of three score years and ten he would surely have been among the first Nobel Prize winners at the beginning of this century. However, a succession of brilliant successors ensured the laboratory's continuing success. For three-quarters of a century, until 1971, every Cavendish Professor was a Nobel Laureate – though not always elected while in office – and so too were many of their colleagues. Maxwell's immediate successor was Lord Rayleigh (1842–1919), a brilliant experimentalist awarded the prize in 1904 for his very accurate measurements of the density of the atmosphere and its constituent gases. This led in 1894 to the discovery of argon and subsequently a whole family of new chemical elements, many of which, such as neon, are a feature of everyday life. In 1884, Rayleigh, wealthy in his own right, retired to continue his research in his private laboratory at Terling Place in Essex. He was one of the last great physicists in the classical tradition. He was followed by J.J. Thomson (1856–1940), already his research colleague, but with a very different outlook, who set in train a whole series of fundamental discoveries in atomic physics. He not only demonstrated experimentally that cathode rays consist of negatively charged particles, which he called electrons, but measured their mass, which proved to be about 1/1000th of that of a hydrogen atom, the lightest atom known. Later he turned his attention to the heavier positively-charged particles, and with F.W. Aston (1877–1945) discovered that neon consisted of atoms of at least two different atomic weights, thus giving rise to the now familiar concept of isotopes.

The First World War interrupted this promising line of research as staff dispersed for the duration to undertake war work of various kinds. Not all returned. Among them was H.G.J. Moseley (1887–1915) who had joined the Cavendish in 1913 after working in Manchester. There he had firmly established experimentally that it was not atomic weight but atomic charge, which he called atomic number, which determined the chemical properties of an element. As Soddy put it, 'Moseley, as it were, called the roll of the elements.' Tragically, he did not live to continue this brilliant research: he was killed at Gallipoli in 1915. Aston went to work at the Royal Aircraft Establishment, Farnborough, and not until 1918 could he start his classic experiments with the famous mass spectrograph with which he investigated the isotopes of over fifty elements. By then Thomson had retired to become Master of Trinity and been succeeded by Ernest Rutherford (Baron Rutherford of Nelson 1871–1937). He had come to Cambridge in 1895 from New Zealand to work with Thomson on the conductivity of electricity in gases. Shortly after, he left to become professor of physics at McGill University, Montreal, and then came back to England, this time to MANCHESTER, where Moseley was one of his colleagues. During these years he evolved his revolutionary concept of atoms consisting of a relatively dense positively charged nucleus surrounded by a cloud of negatively charged electrons sufficient to balance the nuclear charge. Applying the then relatively new quantum theory, the Danish physicist Niels Bohr (1885–1962), working with Rutherford in Manchester, refined the theory by postulating that the electrons moved round the nucleus not at random but in well-defined orbits. The Rutherford/Bohr concept of the atom can fairly be described as the cornerstone of modern atomic physics. For his highly original work Rutherford was awarded a Nobel Prize in 1908 – though his satisfaction at the award was tempered by the fact that it was for chemistry, a science for which he had little regard. In 1919 he was back in Cambridge as Cavendish Professor and head of a truly brilliant research team. Not all of them shared his own taste for experiment. Among them was the eccentric genius Paul Dirac (1902–84), appointed Lucasian Professor of Mathematics in 1933, whose contributions to wave mechanics earned him a Nobel Prize in 1933 (see BRISTOL). Another member of his team was James Chadwick (1891–1974), remembered for his discovery of a third atomic particle, the neutron: for this he was awarded a Nobel Prize in 1935. In that year he left to go to LIVERPOOL, charged with establishing a new centre for nuclear physics research, a task which he accomplished with distinction. From the public point of view the most dramatic event during Rutherford's time was the famous atom-splitting experiment of J.D. Cockcroft (1897–1967) and E.T.S. Walton (b.1903): they shared a Nobel Prize in 1951. During the Second World War Cockcroft was much concerned with the Anglo-American atom bomb project and afterwards he was a natural choice as the first Director of Britain's atomic energy research laboratory at Harwell. Later (1959) he was an appropriate Master of Churchill College, devoted particularly to the advancement of science and technology. Walton's career was different: in 1934 he returned to Trinity College, Dublin, as professor of natural and experimental philosophy.

Apart from the Nobel Prize, Rutherford gained virtually every honour open to him: a peerage, the Order of Merit, Presidency of the Royal Society, and finally, burial in Westminster Abbey. Surprisingly, for he is the acknowledged father of atomic physics, to his dying day he discounted the possibility of atomic power, describing it as 'moonshine'. Much has been made of the simplicity – 'string and sealing wax' – of the apparatus used in his laboratory, much of which has been preserved but is not readily accessible. Sir Edward Bullard (1907–80), one of his research students who went on to make a name for himself as a geophysicist and as Director of the National Physical Laboratory, certainly had doubts about the virtue of this:

I should be prepared to argue that Rutherford was a disaster. He started the 'something for nothing' tradition which I was brought up in and had some difficulty freeing myself from – the notion that research can always be done on the cheap . . . It is wrong. The war taught us differently. If you want quick and effective results you must put the money in.

Post-war developments lend colour to this. Progress in atomic physics increasingly depended on the availability of large, complex and exceedingly expensive particle accelerators and other apparatus and inevitably in this field the initiative passed to the USA who could 'put the money in'. This is reflected in a change in the pattern of research at the Cavendish where today the emphasis is on solid-state physics.

Rutherford's successor was Sir Lawrence Bragg (1890–1971) who, with his father Sir William Bragg (1862–1942) had done pioneer research in crystallography, using the diffraction of X-rays to locate the position of atoms in a crystal lattice: together they had shared a Nobel Prize in 1915. Bragg's own research was concerned mainly with metals and inorganic substances but new techniques made it possible to turn attention to the far more complicated structures of large organic molecules. Thus was born molecular biology, which in half a century has become perhaps the most important field of science, with particular reference to genetics. Here the Cavendish produced stars as brilliant as any in the field of atomic physics. In 1947 John Kendrew (b.1917) and Max Perutz (b.1914) set up a Medical Research Council Unit of Molecular Biology. Together, in 1962, they shared a Nobel Prize for chemistry for their determination of the structure of haemoglobin – the oxygen-transporting pigment of the blood – and the related myoglobin. In the same year the unit moved from the Cavendish to a new Laboratory of Molecular Biology in Hills Road, near Addenbrooke's Hospital. Kendrew remained there until going to Heidelberg in 1975 as Director-General of the newly founded European Molecular Biology Laboratory.

Initially Kendrew and Perutz worked alone at the Cavendish but they were soon joined by two other research workers who were to become equally well known. They were the physicist Francis Crick (b.1916) and a young American biologist James Watson (b.1928). Together they sought to unravel the structure of DNA, the fundamental genetic material of almost all living organisms. In 1943 they announced the famous double helix structure, resembling a ladder twisted into a spiral. This has been acclaimed as one of the greatest biological developments of this century. Together they were awarded a Nobel Prize for Physiology or Medicine in 1962, together with their collaborator Maurice Wilkins (b.1916) who had independently reached a similar result at King's College, LONDON. Thus, in the *annus mirabilis* 1962, the Cavendish boasted two Nobel Laureates for Chemistry and two for Physiology or Medicine. In 1977 Crick moved to the Salk Institute in America but a brass helix on his former home at 20 Portugal Place is a reminder of this dramatic incident in the history of science in Cambridge.

Bragg was followed by a graduate of St John's, Nevill Mott (b.1905) who was Cavendish Professor 1947–71. He had previously done important research at BRISTOL on the structure of metals at the atomic level. He, too, was destined to be a Nobel Laureate, but such are the vagaries of the award that this was not until 1977, some years after he had retired. Against this, however, Brian Josephson (b.1940) was awarded a prize at the early age of 33, for his discovery of the Josephson Effect when he was only 22 and still a research student at Cambridge.

In Mott's time an important new research group emerged. During the war Antony Hewish (b.1924) and Martin Ryle (1918–84) had worked together on radar: afterwards they joined forces at Cambridge to study a part of the celestial spectrum – that represented by radio waves – which had just begun to be investigated in the 1930s. Apart from them, the other important group in Britain was Bernard Lovell's at MANCHESTER. In place of Lovell's huge steerable bowl they built an extended array of dipoles on the ground, giving much finer resolution. With this they detected and catalogued a whole range of 'radio stars'. In 1967, the first pulsar was detected: these are radio stars which 'twinkle' like visible stars. They were later identified as neutron stars, immensely dense bodies created when a massive star collapses under its own weight and the protons and electrons combine to form neutrons.

While the Cavendish Laboratory with its brilliant run of successes dominated Cambridge science for a century, outstanding results have been achieved elsewhere over a broad spectrum. Frederick Gowland Hopkins (1861–1947) brought international fame to the Dunn Institute of Biochemistry by his pioneer work on vitamins from the 1920s. Later, in the post-war years, Alexander Todd (Baron Todd of Trumpington, b.1907) was continuing chemical research on natural products which he had begun as a young research student at OXFORD. He came to Cambridge in 1944 from MANCHESTER, as head of the department of chemistry. With an able team he achieved some outstanding successes in a variety of fields, including vitamins, pigments, nucleic acids, antibiotics and enzymes. Like Rutherford earlier, he gained every possible major distinction: a Nobel Prize (1957), a peerage, the Order of Merit, Presidency of the Royal Society, Master of Christ's College. A colleague, Ronald Norrish (1897–1978) also achieved fame, for his highly original research on the chemistry of very fast chemical reactions. This work was done with George Porter (b.1920) with whom he shared a Nobel Prize in 1967. Norrish remained a dedicated Cambridge man but Porter moved on to other things, including a peerage and Presidency of the Royal Society (see LONDON). When he came to Cambridge, Todd inherited an old-fashioned laboratory in Lensfield Road: in its place there now stands a magnificent new building.

Gowland Hopkins and Todd may be said to have bridged the gap between chemistry and biology. More firmly in the biological camp were Alan Hodgkin (b.1914) and Andrew Huxley (b.1917). Together, encouraged by Adrian, in the post-war years they investigated the way in which nerve impulses are transmitted along nerve fibres: their results were of great fundamental significance and valuable in medical practice. They, too, became Nobel Laureates, in 1963.

People find their way to Cambridge by many different paths, but few more unusual than that taken by Frank Whittle (b.1907). He began his career as an apprentice in the RAF and was awarded a cadetship at Cranwell. There he became interested in aeroplane engines and realized that the conventional engine, with its reciprocating pistons, had in-built limits. By contrast, a gas turbine, burning fuel continuously, could produce much higher speeds and function efficiently at high altitudes. The Air Ministry expressed no interest in developing such a power unit but in 1934 the RAF sent him to Peterhouse to study mechanical engineering: in 1936 he graduated with first-class honours. By then the threat of war was looming and superiority in the air was recognized as a matter of crucial national importance. Whittle was then attached to Power Jets, a company he had formed in 1935 to develop jet propulsion. Even then, it was 1939 before the Air Ministry would provide an aircraft in which the new engine could be tested. By then Hans von Ohain's jet-propelled Heinkel had successfully flown in Germany on 27 August 1939, literally on the eve of war. In the event, however, jet propulsion did not affect the outcome of the war: the Gloster Meteor and the Messerschmitt 262 both came into service, in small numbers, only in 1944. But the consequences for post-war civil aviation were, of course, enormous.

Karl Baedeker's advice to travellers has been much quoted: 'Oxford is on the whole more attractive than Cambridge to the ordinary visitor and the traveller is therefore recommended to visit Cambridge first or to omit it altogether if he cannot visit both.' This advice is dubious at best, and certainly the traveller who wants to visit one of the world's great power-houses of science should on no account omit Cambridge.

CAMELFORD, Corn.

The North Cornwall Museum and Gallery is housed in an old coach-building workshop. Although primarily devoted to illustrating life in North Cornwall over the past century, its exhibits are typical of a much wider area. They include the working tools and equipment of such varied craftsmen as the printer, cobbler, quarryman and cooper. The Domestic Gallery includes a collection of early vacuum cleaners.

CANTERBURY, Kent

The Canterbury and Whitstable Railway was opened in 1830, and was the first in the country to provide a regular passenger service. Trains were hauled over a part of the route by a stationary steam engine, but over the rest by George Stephenson's *Invicta*, which is now displayed in Dane John Gardens.

The West Gate Museum displays arms and armour.

Among famous pupils at the ancient King's School (founded about AD 600) were Thomas Linacre (1460–1524), born in Canterbury and founder of the Royal College of Physicians of London, and William Harvey (1578–1657), born in Folkestone and discoverer of the circulation of the blood.

CARDIFF, W Glam.

The principal city of Wales, with its fortunes long based on coal. The docks were founded by the 2nd Marquess of Bute in 1839 and for many years up to the Second World War Cardiff was the world's greatest coal-exporting port. The Welsh Industrial and Maritime Museum, in the dockland area, presents a comprehensive picture of Welsh industrial and maritime history. The Hall of Power illustrates water, steam and gas power, as well as jet engines. There are also galleries representing transport, and railways and shipping. Adjacent to it is Techniquest, a 'hands-on' science centre.

The Welsh Folk Museum at St Fagans displays the life and culture of Wales in an open-air setting. Many exhibits illustrate crafts and industries – coopering, iron-forging, flour milling, tanning and many others.

In Llandaff Cathedral is a memorial to William Conybeare (1787–1857) who was Dean from 1845. A keen amateur geologist, he is remembered for his reconstruction of skeletons of Plesiosaurus and Ichthyosaurus.

On 15 June 1910 the British Antarctic Expedition, led by Captain Robert Falcon Scott, sailed from Cardiff – where much support had been forthcoming from the local business community – in the *Terra Nova*. With four companions he reached the South Pole on 17 January 1912, only to find that the Norwegian explorer Roald Amundsen had anticipated them by just one month. All five perished tragically on the return journey. The *Terra Nova* and the surviving crew members returned to Cardiff on 16 July 1913. There is a commemorative plaque in the Royal Hamadryad Hospital.

Cardiff was the birthplace (1940) of the physicist Brian David Josephson, inventor of tunnelling superconductors: he was awarded a Nobel Prize in 1970. (See CAMBRIDGE.)

At Lavernock Point (6 miles S) Marconi transmitted the first radio signals over water, to Flat Holm Island in the Bristol Channel (3½ miles). A commemorative plaque in the church of St Lawrence, Lavernock, records the event, which excited world-wide interest.

At St Nicholas is Tinkinswood, a large chamber barrow, notable for including probably the largest capstone in Britain – estimated at 30 tons.

CARDIGAN, Dyfed

Birthplace of Edward Lhuyd (1660–1709), botanist, who was appointed Keeper of the Ashmolean Museum, OXFORD, in 1691. He wrote a comprehensive two-volume natural history of Wales and left an unfinished *Archaeologia Britannica*, containing the first comparative study of the Celtic languages.

CARISBROOKE, IoW

In Carisbrooke Castle, rebuilt in 1587, a working donkey wheel still raises water from a 160-ft deep well. See also ROTHERFIELD GREYS.

CARLISLE, Cumb.

Although much of the material in the Carlisle Museum and Art Gallery is of regional interest there are many industrial exhibits of a more general nature. They include 18th- and 19th-century pottery from many parts of Great Britain, textile machinery, agricultural implements and photographic memorabilia.

CARLOW, Carl.

The dolmen at Mount Browne is one of the largest in Europe. The huge capstone weighs about 100 tons.

Carlow County Museum has a considerable collection of agricultural interest. Exhibits include a blacksmith's forge, a reconstructed dairy and kitchen, and agricultural machinery.

CARMARTHEN, Dyfed

Timothy Richard Lewis (1841–86) was born in Llangain (3 miles S) and educated at the Grammar School in Narberth (15 miles W) before being apprenticed to a local pharmacist. Later he qualified in medicine in ABERDEEN and then joined the Army Medical School, and was sent to India. There he made important discoveries in elucidating the role of the filaria worm in various tropical diseases such as elephantiasis and river blindness.

CARNANEE, Ant.

The 18th-century forge (10 miles E of Antrim) is powered by water and still produces spades and similar items.

CARNFORTH, Lancs.

Although parts are older, Leighton Hall dates mostly from the 18th century. On view is an excellent collection of 18th- and 19th-century furniture made in Lancaster by the local firm of Gillow, with which there was a family connection.

The Steamtown Railway Museum covers nearly thirty acres. It comprises the whole of the original station and depot, complete with turntable, signal box, water towers, cooling plant and so on. There is also a considerable collection of steam locomotives and rolling stock, with which a passenger service is occasionally operated. Included in the collection are the famous *Flying Scotsman* and *Sir Nigel Gresley*.

CARROWMORE, Sligo

Here, some 3 miles W of Sligo, is the largest cemetery of megalithic tombs in Ireland. The most impressive is Maeve's Cairn at Knocknarra.

CASTLE CARY, Som.

Has been a market town since the 17th century and developed a small industry concerned with the making of rope, webbing and twine. These crafts are illustrated in the Castle Cary Museum, with emphasis on the 19th and 20th centuries.

CASTLE DONINGTON, Derbs.

The East Midlands Aeropark covers some 10 acres on the edge of East Midlands Airport. There is a small indoor museum illustrating the history of flight and in the open a collection of historic aircraft, including an Avro Vulcan bomber.

CASTLETOWN, IoM

The Nautical Museum (Bridge St) illustrates the maritime history of the Isle of Man, centred on the yacht *Peggy* (1791), one of a number of vessels built by the Quayle family. Exhibits include photographs and models of local ships and a reconstruction of a sailmaker's loft.

The site of an important Viking ship burial is at Balladoole (1 mile W). It yielded a rich collection of grave goods, now displayed in the Manx Museum, Douglas.

CASTLEWELLAN, Down

The cashel at Drumena (2 miles SW) has been well-restored. The walls are 3 m thick at the base and survive to a height of 3 m (cf STAIGUE). Within is a souterrain 15 m long (cf NEWBIGGING).

CASTEL, Ch. Is.

The Guernsey Folk Museum (National Trust of Guernsey) depicts various aspects of rural industry, including agriculture, cider-making and quarrying.

CELBRIDGE, Kild.

Castletown House, built in the Palladian style, is the largest house in Ireland, dating from the early 18th century. At one time more than a hundred servants were employed. It can now be seen very much as it was in its heyday, with all the accoutrements of a grand mansion.

Celbridge Motor Museum displays a collection of vintage and veteran motor-cycles and cars, mostly Irish.

CENARTH FALLS, Dyfed

The Fishing Museum has an exceptional collection of exhibits illustrating the history of rod-and-line fishing. Of particular interest are the coracles, leather boats for which the River Teifi is famous.

CHALFONT ST GILES, Bucks.

The Chiltern Open Air Museum has acquired and re-erected many buildings of historic interest, mainly agricultural. Exhibits include a furniture factory and an early 18th-century toll-house from HIGH WYCOMBE.

CHARD, Som.

Chard has a thin claim to be regarded as the birthplace of the aviation industry. In the late 1830s two local lace manufacturers, W.S. Henson (1805–88) and John Stringfellow (1799–1883), inspired by the aerodynamic principles laid down by Sir George Cayley (1773–1857), took a serious interest in heavier-than-air flight. In 1842 Henson designed and patented a specification for an Aerial Steam Carriage. This was a monoplane with a 150-ft wingspan, powered by a steam engine driving two propellers. It anticipated later designs in having a tail-unit to provide directional control and stability and a tricycle undercarriage. With Stringfellow he conceived an ambitious scheme for an Aerial Transit Company. The venture was doomed to failure as the power:weight ratio of the best available steam engines was insufficient for sustained flight. Henson emigrated to America but Stringfellow persevered. A smaller machine, taking off from a guide-wire, is alleged to have made a short flight in 1848. In 1868 he exhibited a steam-driven model triplane at the first Aeronautical Exhibition at the Crystal Palace. His work, original if unsuccessful, is commemorated in the Chard and District Museum.

A water-mill at Hornsbury can be traced back to the 14th century. The present overshot wheel dates from 1870, and the mill's equipment and machinery is intact.

Hill's Plumbing Museum is unusual – possibly unique – in being devoted entirely to equipment used in plumbing.

CHARTERHOUSE, Som.

With Derbyshire and Wales, Mendip was the great lead-mining area of Britain and it has been worked there from Roman times until very recently. The area around Charterhouse shows abundant surface evidence of this activity: mineshafts, slag heaps and condensation flues.

CHATHAM, Kent

The Chatham Historic Dockyard Museum displays, on a grand scale, four centuries of naval history. Exhibits include demonstrations of rope-making, sail-making and ship restoration. Nelson's *Victory* was laid down here in 1759: the last vessel built was a submarine, *Okanagan*, for the Royal Canadian Navy. The dockyard closed in 1984.

The Royal Engineers Museum (Brompton Barracks) depicts all aspects of military engineering, from Roman times.

CHEDDAR, Som.

The famous Gorge contains three large caves which show evidence of human occupation in the Palaeolithic age (about 12,000 BC). The best-known is Gough's Cave, discovered in 1893. There is a small museum on the site.

CHEDWORTH, Glos.

A fine example of a Romano-British villa, extensively excavated to show hypocausts and mosaics.

CHELMSFORD, Esx

Terling Place is the family seat of the Rayleigh family. J.W. Strutt (1842–1919) became the 3rd Lord Rayleigh in 1873. He achieved fame as a mathematical physicist and in 1904 shared the Nobel Prize for Chemistry with Sir William Ramsay (1852–1916) for their researches on the rare gases of the atmosphere. In 1879 he was appointed Cavendish Professor of Physics at CAMBRIDGE but from 1884 did most of his research in a private laboratory at his home.

CHELTENHAM, Glos.

Birthplace of Sir Rowland Biffen (1874–1949) noted for his use of genetic principles to improve cultivated plants. First Director of the Plant Breeding Institute, CAMBRIDGE (1912).

CHEPSTOW, Gwent

Chepstow Museum is housed in Gwy House, dating from about 1790, and is concerned largely with the history and development of Chepstow and its locality. The 'Chepstow at Work' exhibition depicts local trades and industries, including shipbuilding, engineering, salmon-fishing and brush-making.

The museum also displays items recovered during excavations at Caerwent (Venta Silurum) (5 miles SW), the tribal capital of the Silures. Parts of the walls survive to a height of 5 m; internally the foundations of an octagonal temple and some houses have been uncovered.

CHESTER, Ches.

Famous in the 12th century for its school of natural philosophers. Its best-known member was Robert of Chester (born early in the 12th century) remembered for his *Book of the Composition of Alchemy* (1144), a translation into Latin of an 8th-century Arabic text. He made the first Latin translation of the Koran (1143). He also translated an Arabic text on algebra and was the first to use the word sine (*sinus*) in its trigonometrical sense. The mathematician John Wallis (1614–72), a founder of the Royal Society, was made Bishop of Chester in 1668.

Exhibits at the Water Tower Museum include a camera obscura.

CHESTERFIELD, Derbs.

The centre of Chesterfield is dominated by two disparate buildings: Trinity Church, with its famous twisted spire, and the factory of Robinson and Sons, an old established textile manufacturer with a special interest in surgical dressings and related products. The railway pioneer George Stephenson (1781–1848) spent the last year of his life at Tapton Hall and is buried in Trinity Church beneath the communion table. His executor was Josiah Robinson who received a gold tie-ring – still a treasured family possession – in recognition of his services. The most distinguished member of the Robinson family was the organic chemist Sir Robert Robinson (1886–1975), famous for his research on natural products. In his long life he gained virtually all the honours open to him: President of the Royal Society (1945–50), Nobel Prize for Chemistry (1947), Order of Merit (1949), and innumerable honorary degrees.

CHICHESTER, W Ssx

Suddenly famous (1994) for the discovery of Boxgrove Man, hailed as the earliest European. He died some four million years ago and his remains – a shinbone and later an incisor tooth – were discovered deep in a huge gravel pit at Boxgrove (5 miles E).

The Weald and Downland Open Air Museum at Singleton (10 miles N) displays several reconstructed historic buildings, including a toll-house from Shoreham. Exhibits include charcoal burning, once vital for the ironworks of the Weald; a working water-mill for grinding flour; and a windpump.

CHILD OKEFORD, Dors. See BLANDFORD FORUM

CHILLENDEN, Kent

Chillenden Windmill is a good example of a post mill, built in 1868. It is no longer operational but most of the machinery and equipment survives.

CHIPPING CAMDEN, Glos.

The Woolstaplers Hall Museum displays a variety of material of technical interest –

instruments, typewriters, vacuum cleaners, apothecaries' materials, photographic material, etc.

CHIPPING NORTON, Oxon.

The Rollright Stones (2 miles NW) – known locally as the King's Men – is one of the best-known stone circles. It now comprises seventy-seven severely weathered limestone blocks but some may be fragments of larger ones. It dates from about 2000 BC.

CHIRBURY, Shrops.

Mitchell's Fold, dating from the Bronze Age, is a slightly oblate ring of sixteen stones, with an 'altar' stone 100 m SW.

CHITTLEHAMPTON, Devon

Modern military technology is exemplified at the Cobbaton Combat Museum, which displays some thirty British and Canadian armoured fighting vehicles – tanks, trucks and cars – of the Second World War. They are maintained in running order.

CHRISTCHURCH, Dors.

The Christchurch Tricycle Museum (The Quay) is devoted to multi-wheeled cycles, including a pentacycle.

CHURCHILL, Oxon.

Birthplace of William Smith (1769–1839), the 'Father of English Geology'. As a superintendent of canal construction he travelled all over the country and could examine the exposed strata at first hand. His *Delineation of the Strata of England and Wales* was published in 1815. There is a memorial to him on the village green.

CHURCH STRETTON, Shrops.

Acton Scott Farm Museum demonstrates farming practice at the turn of the century, using cart horses and machinery.

CIRENCESTER, Glos.

Cirencester (Corinium) on the Fosse Way (see ROMAN ROADS) was founded in the 1st century AD and, after London, became the principal Roman city in Britain. Substantial remains of the amphitheatre and fragments of the town wall survive. The Corinium Museum has an outstanding collection of material relating to Roman Britain, including several fine mosaics.

CLESTRAIN, Ork.

The megalithic tomb Maes Howe is one of the finest in the British Isles, 40 m in diameter and 3.5 m high. The entrance passage – so oriented that the rising sun shines down it at mid-winter – is 10 m long.

CLIFTON, Cumb.

Birthplace of John Wilkinson (1728–1808), ironmaster. He pioneered the use of iron for a wide variety of purposes, from coffins to canal boats. His most important contribution, however, was his boring engine (1774) first used for making cannon but later of crucial importance for boring the huge cylinders needed for Boulton and Watt steam engines.

CLITHEROE, Lancs.

Clitheroe Castle Museum is devoted primarily to local life and consequently has some exhibits not generally seen in other parts of the country. One is a reconstructed clogger's workshop: wooden clogs were the common footwear in the textile mills. There is also a reconstructed printer's workshop.

CLOGHER, Tyr.

Knockmany Passage Grave is of particular interest because of its carved patterns of circles, zigzags and spirals.

CLUNY, Fife

Birthplace of J.T.R. Macleod (1876–1935) who with F.G. Banting introduced the use of insulin for the treatment of diabetes in Canada in 1922. They shared a Nobel Prize in 1923. (See ABERDEEN.)

CLYDEBANK, Strath.

As its name implies, the Clydebank District Museum, in the Old Town Hall, is devoted mostly to exhibits of local interest, including ship models. Additionally, it displays a collection of historic sewing machines.

COATBRIDGE, Strath.

The Summerlee Heritage Trust is housed in a large disused crane factory. The exhibits are devoted to the history of the Scottish iron, steel and heavy engineering industries. Much of the machinery is operational, driven from line shafting. Exhibits include a winding engine, brought from Cardowan Colliery, steam cranes, and steam locomotives.

COCKERMOUTH, Cumb. See EAGLESFIELD

COCKSBURNSPATH, Loth.

Siccar Point is famous in geological history for its striking nonconformity, first noted by James Hutton (1726–97) and referred to in his *Theory of the Earth* (1788). His views were radical: while Bible literalists accepted the age of the earth as being around 6,000 years, Hutton believed it to exist in perpetuity – 'no vestige of a beginning, no prospect of an end'.

COLCHESTER, Esx

As Roman Camulodunum, this is reputedly the oldest town in England. Birthplace of William Gilbert (1544–1603), physician to Elizabeth I and James I & VI. Remembered as an original investigator of magnetism and electricity: his *De Magnete* (1600) is one of the great classics of science. Dryden proclaimed that 'Gilbert shall live till loadstones cease to draw'. He was buried there and there is a monument in Holy Trinity Church. Also the birthplace of Francis Hauksbee (*c.* 1666–1713), remembered for the design and construction of air-pumps, and experimental work with them. He showed that at low pressures air glows when excited by an electric discharge. His *Physico-Mechanical Experiments* first appeared in 1702 and was translated into French and Italian. Grandad's Photography Museum (East Hill) displays some 400 cameras (from 1850), magic lanterns, darkroom equipment, and other photographic memorabilia. Bourne Mill (NT), a two-storeyed water-mill for grinding corn, dates from 1591.

COLEFORD, Glos. See FOREST OF DEAN.

COLERAINE, Londy

The well-preserved souterrain at Dunalis (3 miles SW) consists of three passages on different levels. Some of the lintels bear ogam inscriptions (cf NEWBIGGING).

COLSTERWORTH, Lincs.

In his youth Isaac Newton (1642–1727) had a passion for building working models such as windmills and waterwheels, as well as simple measuring devices such as sundials. A dial mounted on Colsterworth church was reputedly carved by Newton at the age of nine. (See GRANTHAM.)

CONISTON, Cumb.

A scenic attraction for visitors, Coniston was once the centre of an important copper-mining area. The Copper Mines Valley is full of evidence of old workings and a number of buildings survive. Mining began in the 16th century but was most active in the middle of the 19th. The Church Beck provided water power.

CONWY, Gynd

The suspension bridge across the River Conwy, built in 1826, is a monument to Thomas Telford (1757–1834) and part of his grand scheme to improve the London–Holyhead road. A road tunnel beneath the river was opened in 1991. On Penmaenmawr headland is the Druid's Circle (4 miles SSW) a large Neolithic stone circle. In 1695 Edward Lhuyd (1660–1709), keeper of the Ashmolean Museum in OXFORD, described it as 'the most remarkable monument in all Snowdon'. Nearby Craig Lwyd is the site of a stone-age factory from which axes were traded as far as southern England. At Llansantffraid Glan Conwy is Felin Isaf Water Mill, a fully restored flour mill in which much of the machinery dates back to 1730.

COOKSTOWN, Tyr.

The Wellbrook Beetling Mill (1765) (3 miles W on A505) used water-powered hammers to give the fabric a sheen in the last stages of linen processing. The water-wheel is somewhat unusual in being breastshot.

CORBRIDGE, Northld

The Roman fort at Corbridge (Coriosopitum), just south of HADRIAN'S WALL, has been partially excavated, but most of what is visible dates from rebuilding in the 3rd and 4th centuries. The outstanding feature is the headquarters building (*principia*). There is a good site museum.

CORK, Cork

Busy commercial seaport: third largest city in the Republic of Ireland. The University College, originally Queen's College (1845), was founded in 1908 as a constituent college of the National University of Ireland. The city's most famous scientist was George Boole (1815–64), famous for his work in mathematical logic and founder of Boolean algebra. Robert Boyle (1627–91), a founder member of the Royal Society, widely known as 'the father of chemistry' was the seventh son of the first Earl of Cork.

CORNHILL-ON-TWEED, Northld

The original Heatherslaw Mill, on the River Till, dates back to the 13th century but the present building dates only to the mid-19th century. It comprises two mills. One is operational, grinding flour; the other has been partly dismantled to display its internal workings.

CORRIMONY, Hghld

A well-preserved passage grave – possibly used also as a shrine – is surrounded by a circle of standing stones.

CORRIS, Gynd

Mining village in the Dulas Valley north of MACHYNLLETH, famous for its high-quality blue slate. Over the years some 20 quarries and mines were opened, some dating back to medieval times, but only one is now operational. Slate not used locally was exported by sea from Derwenlas, on the Dovey Estuary. The 1858 Tramroad Act made it possible to build the 11-mile narrow-gauge Corris, Machynlleth, and River Dovey Tramroad. Initially, horse traction was obligatory but in 1878 steam locomotives were introduced and it became the Corris Railway. For many years it prospered, carrying both freight – 17,000 tons of slate in 1902 – and (from 1883) passengers, but business declined after the First World War and it was closed in 1948. A local preservation society has been working for many years to reopen a section of the line. There is a small Railway Museum at the old Corris Station. At the local Corris Craft Centre some of the underground slate workings have been restored and opened to visitors. Although essentially a tourist attraction – based on the Arthurian Legend – it provides an excellent opportunity to see working conditions in the past and some of the enormous caverns excavated underground using only hand tools and blasting with gunpowder.

CORSHAM, Wilts.

Bath stone (oolitic limestone) has long been a prized building material and the underground Pickwick Quarry is an important source. The Bath Stone Quarry Museum illustrates the history of Bath stone from its extraction to the finished product. Visitors can tour the underground workings to inspect equipment and see how it was used. There are demonstrations of the way in which the stone is dressed for use.

COTEHELE, Corn.

The Cotehele Estate (NT) was originally a self-sustaining medieval community on the River Tamar. A warehouse on Cotehele Quay serves as a maritime museum and includes a restored Tamar sailing barge. An 18th-century grain mill is powered by a large overshot water-wheel.

COTTESMORE, Leics.

The Rutland Railway Museum, on the Ironstone Sidings, displays a collection of industrial locomotives and rolling stock, and a steam crane.

COULTERSHAW, W Ssx

Coultershaw Beam Pump was installed in 1790 to pump water from the River Rother to Petworth. It fell into disrepair but has now been restored to working order by a local conservation group and supplies a fountain.

COVENTRY, W Mids

Coventry, an important centre for the textile industry since the 17th century, has been identified with Middlemarch, the Midlands community of George Eliot's novel. In 1904 this took a new turn with the start of rayon manufacture by Courtaulds. (See HALSTEAD.)

Today, it is more generally identified with light engineering. Since 1870, bicycles, and later motor-cycles, have been made there. In 1888 this led J.B. Dunlop (1840–1921) to move his pneumatic tyre factory there – at that time the bicycle industry was his main customer. The Singer Sewing Machine works was built in Canterbury Street about 1880 but from 1904 began to make Singer motor-cars.

The Midland Air Museum, at Coventry Airport, displays aircraft from freighters to fighters. There is also an exhibition of memorabilia relating to Sir Frank Whittle, who was born in Coventry. The Museum of British Road Transport includes motor-cycles and commercial vehicles as well as motor-cars.

The Herbert Art Gallery and Museum has many exhibits illustrating the city's past and present industries – weaving (especially ribbon), watchmaking and sewing machines.

COWDENBEATH, Fife

Birthplace of Sir James Black, pharmacologist. He discovered the beta-blocking drugs for the treatment of certain forms of heart disease, and cimetidine for treating stomach ulcers. Nobel Laureate 1988.

COWSHILL, Dur.

This town lies in an area where lead has long been mined and smelted and there is much surface evidence of old workings. The most remarkable survival is the ore-crushing mill at Killhope (2 miles NW) dominated by a 10-m water-wheel erected in 1876–8 to deal with the exceptionally hard local ore.

CRANBROOK, Kent

Union Mill is a fine example of a smock mill. Built in 1814 it towers 70 ft above the town. It is still operational but powered by an electric motor.

CREED, Corn.

In 1793 William Gregor (1761–1817) – a Fellow of St John's College, Cambridge, and born in the nearby village of Trewarthenick – was appointed Rector of Creed, where he lived until his death. He was a keen amateur chemist and mineralogist and in 1791 published an analysis of a black magnetic sand found some miles away at MANACCAN. From this he obtained the oxide of a new element, for which he proposed the name menacchanine. Two years later it was independently discovered by the German chemist M.H. Klaproth, who called it titanium, the name used today.

CREETOWN, D & G

The Gem-Rock Museum displays a variety of natural minerals, including local beach agate. There is also a workshop, demonstrating the way in which gemstones are cut and polished.

CREEVYKEEL, Sligo

A fine example of a megalithic court cairn is situated here. A short narrow entrance passage leads to a circular court from which burial chambers open out.

CREGNEASH, IoM

Cregneash Folk Museum illustrates life in a typical Manx crofting/fishing village by means of reconstructed workshops and cottages.

CROMARTY, Hghld

Birthplace of the geologist Hugh Miller (1802–56) famous for his researches on the Old Red Sandstone. Hugh Miller's Cottage contains memorabilia and geological specimens.

CROMFORD, Derbs.

This manufacturing town was at the very heart of the Industrial Revolution, for it was here that Richard Arkwright (1732–92) – a pioneer of the mechanical spinning of cotton – set up his water-powered mill in 1771. It still stands, though reduced by a fire in 1929. A little to the north is his elegant Masson Mill which he built 1784–5: the date on its front refers to the founding of his first business in Nottingham in 1769. He is buried at St Mary's Church. Hand-weaving continued as a cottage industry and the weavers' cottages in North Street are characterized – like their counterparts elsewhere – by the long 'weaver's' windows designed to throw maximum light on the looms.

The Good Luck Lead Mine was once the most successful of local lead mines but it has long been closed. Substantial surface buildings remain, however, and visitors are able to tour part of the extensive underground workings.

CROSBY RAVENSWORTH, Cumb.

Castlehouse Scar is a small stone circle (7 m diameter) but unusual in that all its original eleven stones remain standing.

CROSSMAGLEN, Ant.

At Anaghmore (1½ miles N) is a good example of a court grave.

CRUMLIN, Gwent

The iron railway viaduct, 340 m long and 70 m high, is an outstanding example of Victorian engineering.

CRYNANT, W Glam.

The Cefn Coed Mining Museum displays machinery and techniques – including a steam winding-engine – once used at the Cefn Coed Mine.

CULTRA, Ant.

The Ulster Folk and Transport Museum houses the Irish Railway Collection. The rolling stock includes carriages and wagons and the huge 185-ton *Maeve* locomotive.

CUSHENDALL, Ant.

At Tievebulliagh (3 miles W) there is an outcrop of hard rock, once much prized for making stone axes, which were traded far afield. Of the workshop itself no trace remains, but the local scree slopes are littered with broken rejects and stone flakes (cf GREAT LANGDALE).

CUSHENDUN, Ant.

The megalithic tomb on Mt Carranmore (5 miles NW) is a good example of a passage grave. The large burial chamber has a corbelled roof.

D

DALMELLINGTON, Strath.

The Scottish Industrial Railway Centre at Minnivey Colliery exhibits rolling stock, cranes, and steam and diesel locomotives.

DARLINGTON, Dur.

The Darlington Railway Centre and Museum, based on the restored railway station of the Stockton and Darlington Railway, displays a large collection of early locomotives and rolling stock, including the famous *Locomotion*.

Tees Cottage Pumping Station (Coniscliffe Rd) is a water-pumping station dating back to the 1840s. Exhibits include a gas engine, complete with its own gas-generating plant, and a 1903 compound beam engine.

DARTMOUTH, Devon

The two great names in the early history of the steam engine are Thomas Savery (1650–1715) and Thomas Newcomen (1663–1729). The two worked quite independently but by chance both were born within a dozen miles of each other, Newcomen in Dartmouth and Savery in Modbury (12 miles E). In the event, Savery's engine was not a success but his patents were far-reaching and Newcomen had to make an accommodation with him. The coincidence is not perhaps as great as it may seem because both were brought up in the Cornish mining community where pumping water – the first major application of steam – from mines was a constant problem.

The Newcomen Engine House (adjacent Butterwalk) displays an original Newcomen beam engine dating from about 1725.

DEAL, Kent

The Time Ball Tower (Victoria Parade) has a semaphore tower, relic of an old system of mechanical telegraphs introduced by the Admiralty during the Napoleonic Wars to connect London with South Coast ports.

DENBY, Clwyd

The Fechan Valley is rich in sites of archaeological interest and many were identified and restored when the Llyn Brenig reservoir was built in the 1970s. From the car park two trails give easy access to many of the sites.

DENT, N Yorks.

Birthplace of Adam Sedgwick (1785–1873) whose father was vicar of Dent: he went to school nearby at Sedbergh. Founder of the Cambrian System of rocks. For 55 years he was

Professor of Geology at CAMBRIDGE, where his lasting memorial is the Sedgwick Museum of Geology. A fountain set in a massive piece of granite is a memorial to him here. He is also commemorated by a wall plaque and stained-glass window in the village church.

DERBY, Derbs.

Industrial city on the River Derwent, taken by the Danes in 880 and renamed Deoraby: now an important manufacturing centre. The Silk Mill, the first textile factory in Britain, was built in 1717 by Thomas Lombe (1685–1739). The present building is lower than the original as the result of a disastrous fire in 1910. It houses the Derby Industrial Museum, which displays much material relating to Derbyshire's industrial history, including Rolls Royce engines made in the famous local works. Derby is also an important centre of porcelain manufacture, initially by the Duesbury family 1756–1814.

It was the home-town of John Flamsteed (1646–1719), first Astronomer Royal, and of Herbert Spencer (1820–1903), evolutionary philosopher, reformer and sociologist who had his early schooling there. John Whitehurst (1713–88), best-known as a geologist, founded in Derby in 1736 a famous workshop for the manufacture of watches and scientific instruments.

But perhaps the most widely known citizen is the painter Joseph Wright (1734–97) who was born there and did much of his work there. He is noted for his effective portrayal of the effects of light. Some of his paintings had scientific themes: well-known examples are *The Orrery* (1766) and *Experiment with the Air Pump*. He also painted industrial scenes such as *Blast Furnace by Moonlight* and *An Iron Forge*. Examples of his work are to be seen in the Derby Art Gallery.

Derby was also the home-town of Sir George Simpson (1878–1965), meteorologist, who was educated at the Diocesan School. He was Director of the Meteorological Office 1920–38.

The eccentric chemist Henry Cavendish (1731–1810), who first identified hydrogen ('inflammable air') as a chemical entity, is buried at the Cathedral of All Saints. Nephew of the 3rd Duke of Devonshire (see BAKEWELL) he was termed 'the richest of the learned and the most learned of the rich'.

DERRY, Dngl

The massive cashel (fortified homestead) at Grianiàn of Ailach was once the seat of the Kings of Ulster. Some restoration has been done and the walls rise to a height of 5 m (cf STAIGUE).

DEVIZES, Wilts.

The Kennet and Avon Canal, completed in 1810, included a flight of twenty-nine locks at Devizes. There is an exhibition illustrating the history of the canal at the Kennet and Avon Canal Centre, Couch Lane.

DIDCOT, Oxon.

The Didcot Railway Centre has a large collection of locomotives, rolling stock and much other material relevant to the history of the Great Western Railway.

DINGLE, Kerry

The Dingle Peninsula is rich in sites of archaeological interest. At Ballintaggart are two ogam stones inscribed with runes, the twenty-letter alphabet used around the 4th century AD for writing in Irish and Pictish. At Ventry (3 miles W) is a well-preserved Iron Age promontory fort.

DOBWALLS, Corn.

Archibald Thorburn (1860–1935) is recognized as one of the greatest wildlife painters of all times, and the meticulous accuracy of his paintings gives him a scientific as well as artistic interest. His greatest achievement was the monumental *Birds of the British Isles* (1885–97), depicting over 400 of about 470 recognized species. Some two hundred of his paintings are displayed at the Thorburn Museum and Gallery.

DOLGELLAU, Gynd

Welsh market town in an area rich in minerals, including copper, lead and zinc. It is, however, most famous for its gold mines, located in the Mawddach Valley. The most famous were at Gwynfyndd and at Clogau, on the Barmouth road. Only trifling amounts are now extracted – though visitors pan specks from the river gravel – but since the 1850s over 130,000 ounces have been mined.

DONARD, Wklw

The Transport Museum Society of Ireland (Castle Ruddery Depot) has a large collection of commercial vehicles and many items relating to the history of transport.

DORCHESTER, Dors.

Maiden Castle is a magnificent hillfort, commenced about 3000 BC and continuing in use until AD 44, when it fell to Vespasian.

Thomas Sydenham (1624–89), 'the English Hippocrates', was born at Wynford Eagle (6 miles NW).

DORKING, Sry

Birthplace of Thomas Malthus FRS (1766–1834), pioneer of population science and economics. His *Essay on the Principles of Population* (1798) is one of the classics in this field and influenced the thinking of Charles Darwin on evolution.

DOUNBY, Ork.

The water-powered corn mill here is unusual in that the shaft is vertical, in the Norse style.

DOUNE, Cent.

The Doune Motor Museum houses a collection of some fifty vintage and later motor-cars, including many famous marques.

DOVER, Kent

The Dover Castle clock is a well-preserved example of an early turret clock with a striking-chain. Originally dated to 1348 it is now believed to be of the mid-17th century.

The Connaught Pumping Station is the responsibility of the Dover Transport Museum Society. Exhibits include a carriage of 1890 taken from the cliff railway at Folkestone, a Fox-Walker locomotive (1878), and an inverted triple-expansion steam engine.

The Roman Dover Tourist Centre (New Street) is based on a Roman villa of about AD 200, discovered in 1971, and the remains of a later Roman fort. There is a small associated museum.

DOWNE, Kent

Downe House was the home of Charles Darwin (1804–82) and is now preserved as a national monument. The study in which he worked for 35 years is still exactly as it was when he died. Here he wrote *Journal of the Voyage of the 'Beagle'* and his great *Origin of Species*, among other major works.

Downe was home from 1932 until his death of the anatomist and anthropologist Sir Arthur Keith (1866–1955), who became Director of the Buckston Browne Research Farm there. His many important publications include *The Antiquity of Man* (1915).

DOWNPATRICK, Down

A one-mile section of the Belfast and County Down Railway has been restored. Working locomotives include the steam-powered *Guinness* and two E-class Maybach diesels.

The Ballynose stone circle dates from the early Bronze Age. It consists of a central mound with a ring of closely spaced uprights.

DRAPERSTOWN, Londy

The Museum at Upperlands displays machinery used in the linen industry.

DRE-FACH FELINDRE, Dyfed. See NEWCASTLE EMLYN

DREGHORN, Strath.

Birthplace of John Boyd Dunlop (1840–1921), inventor of the pneumatic tyre (1888). By profession he was a veterinary surgeon, practising very successfully first in EDINBURGH and later in BELFAST.

DUBLIN, Dub.

Historical differences between Ireland and other parts of Britain – and indeed of western Europe generally – are reflected in the development of science here. Dublin, at the mouth of the River Liffey and capital of the county of the same name, was originally a Viking settlement, dating from the time of the great Scandinavian attacks on western Europe and beyond, between the late 8th and mid-11th centuries. It seems at first to have been no more than a base from which to raid the surrounding countryside, but by the 10th century it began to be a significant commercial and political centre. In 1170 it was uneasily occupied by the Normans and became the capital of the Norman/English colony in Ireland, covering an area – the Pale – somewhat larger than the present county. For a time it flourished, and the Crown asserted its authority in the 16th century, but in the later Middle Ages the Tudors were obliged to reconquer most of Ireland. This task was completed by the beginning of the 17th century but by then the Reformation created new problems. Not until the close of the 17th century, after two civil wars, was English control of Ireland finally established. This conferred a degree of stability favourable to the pursuit of learning, whose lamp had hitherto burned sporadically only in a few monastic institutions. Dublin, as the principal city in Ireland, once again prospered. The beginning of the new era can be dated to 1591 when Trinity College – modelled on Trinity College, CAMBRIDGE – was founded on the site of the disbanded monastery of All Hallows, in the countryside close to the River Liffey. The present buildings date from the mid-18th century.

In the 18th century Dublin became a notable cultural and social centre, possessing many fine buildings located in well-planned streets and open spaces. This Georgian imprint still survives but lacks its original grandeur. The end of the century, however, saw another change of fortune. By the Act of Union of 1801, Dublin ceased to be a parliamentary capital and many leading citizens followed the legislature to London. At that time it was the second largest city in the British Isles but by 1900 it had fallen to tenth. Not until 1922 was a measure of political independence restored to southern Ireland, with the creation of the Republic of Ireland.

With this turbulent background it is scarcely surprising that scientific genius flourished relatively late in Ireland, and then largely in Dublin, nor that when it did it was frequently tempted away by the greater opportunities abroad – especially in England and Scotland and later North America. Indeed, considering the small population, it is remarkable how much was achieved.

For nearly three centuries higher education in Dublin was dominated by Trinity College. The foundation of the Catholic University of Ireland in 1851 reflects the rise of an increasingly influential Roman Catholic middle class anxious to assert itself. Today – as University College, Dublin – it is part of the National University of Ireland, founded in 1908, which has constituent colleges in CORK and GALWAY.

The golden age of science in Dublin was in the 18th and 19th centuries, though an exception might be made for James Ussher (1581–1656), Archbishop of ARMAGH, famous for his calculation that the Creation occurred in 4004 BC, an early example of cosmological prediction. He was a scholar of Trinity College and his portrait hangs in the Examination Hall there. An older generation of chemists would immediately recognize the name of Peter Woulfe (1727–1803), for the two-necked 'Woulfe Bottle' which he invented was, until quite recently, in use in virtually every chemical laboratory in the world. It was designed to remove 'fumes very hurtful to the lungs' by bubbling impure gases through water. Save that he came from Ireland, possibly from Dublin, little is known about his personal life. He settled in London, became friendly with the leading chemists of the day – including Humphry Davy (1778–1829) and Joseph Priestley (1737–1804), to whom he sometimes lent apparatus – and was elected Fellow of the Royal Society in 1767. In 1776 he was appointed to deliver the Society's first Bakerian Lecture, now one of its most prestigious awards. A devout Christian and notable eccentric, he believed in the transmutation of metals long after most of the scientific world had abandoned it: he attributed his own lack of success to the 'want of due preparation by pious and charitable acts'. Davy said that he attached prayers to his apparatus.

Woulfe was the first of a number of distinguished chemists associated with Dublin in the 18th century. Chief among them was William Higgins FRS (1763–1825), born in Sligo. As a young man he worked in London in his uncle's laboratory in Soho and then studied chemistry at Oxford. In 1792 he returned to Ireland to be chemist to the Apothecaries Hall of Ireland and from 1795 to 1822 was chemist to the Irish Linen Board. In 1789, when he was only 26, he published his *Comparative View of the Phlogistic and Antiphlogistic Theories*. This certainly contains the elements of a chemical atomic theory but his claim to have substantively anticipated John Dalton (1766–1844) in this is not now generally accepted. Higgins did his case no good by overstating it and by his quarrelsome disposition. Another Irish chemist of this period, from Galway, was Richard Kirwan (1733–1812) who originally studied law and practised as a barrister, though without much success, in London and Dublin. From 1777 to 1787 he worked in his own private laboratory in London – during which time he was elected a Fellow of the Royal Society – and then returned to Dublin where he was elected President of the Royal Irish Academy. Initially a convinced believer in the Phlogiston Theory of combustion, he was later convinced by Lavoisier's evidence that in fact it involved combination with oxygen. His successful *Elements of Mineralogy* (1784) was the first systematic treatise on the subject. A notable eccentric, he is recorded as having habitually worn an overcoat and hat indoors and subsisting on a diet of ham and milk.

Another important Dublin scientist, though not Irish by birth, was John Brinkley (1763–1835). Born at Woodbridge, Suffolk, he read mathematics at Caius College, Cambridge, and then became assistant to Nevil Maskelyne (1732–1811), Astronomer Royal, at Greenwich. On Maskelyne's recommendation he was appointed Professor of Astronomy at Dublin in 1790 and in 1792 was appointed first Royal Astronomer for Ireland. In the meantime he was ordained a priest and developed in parallel a distinguished ecclesiastical career: when he retired as astronomer in 1826 he became Bishop of Cloyne. In 1808, after acquiring an 8-ft meridian circle, Brinkley embarked on an ambitious programme of observations designed to determine the parallax of the fixed stars, and thus to calculate their distances. In the event, his observations were challenged, and eventually discredited, but they stimulated interest in this field which led to ultimate success by others.

Chemical achievement continued in the 19th century. James Muspratt (1793–1886) was born in Dublin and had an adventurous career. As a young man he fought in the Peninsular War; he enlisted in the Royal Navy as a midshipman and deserted at Swansea by swimming ashore, and eventually made his way back to Dublin. Supported by a small inheritance, he enjoyed to the full the city's literary, dramatic and scientific life. He set up as a small chemical manufacturer and then in 1823 took advantage of the abolition of the £30 per ton duty on salt. This made the Leblanc soda process much more attractive to work and Muspratt moved to LIVERPOOL, close to the Cheshire salt fields, to establish a business there. Despite endless litigation, for the Leblanc process was not environmentally friendly, he was ultimately very successful and he can fairly be described as the founder of the British alkali industry. Among his friends in Dublin was the chemist Sir Robert Kane FRS (1804–90) who was appointed Professor of Chemistry in the Apothecaries Hall in 1831, a post earlier held by William Higgins. His three-volume *Elements of Chemistry* (1841) was acknowledged to be the best and most comprehensive chemical textbook of its day. A later book, *Industrial Resources of Ireland* (1844) attracted the attention of Sir Robert Peel and changed his career: he was appointed adviser to the Irish government on the development of education and industry. He was appointed director of the newly founded Museum of Economic Geology, which evolved into the Royal College of Science of Ireland, in turn absorbed into University College, Dublin, in 1926. He was also President of Queen's College, Cork. Dublin was the birthplace of the American chemist John William Mallet (1832–1912). He studied at Trinity College and Göttingen before settling in America in 1853. He served in the Civil War on the Confederate side and then pursued an academic career, latterly as Professor of Chemistry in Virginia University.

Dublin also provided some distinguished physicists and mathematicians in the 19th century. Sir Robert Ball (1840–1913) was born in Dublin and educated at Trinity College. He became Royal Astronomer for Ireland 1874–92, and then moved to Cambridge as Professor of Astronomy and Geometry. Humphrey Lloyd (1800–81), too, was born in Dublin and was a graduate of Trinity College, of which he eventually became Provost (1867). He is remembered for his work on optics and for his mapping of the earth's magnetic field, especially in Ireland. This was part of a national survey set up by the British Association and led to the discovery of significant temporal variations in the field.

But in this period two Dublin men stand out as being of truly international stature. The first was Sir William Rowan Hamilton (1805–65), one of the few great mathematicians with a gift for involved mental calculations. He was a precocious child who by the age of 13 had mastered as many languages, including Sanskrit, Arabic and Persian. He entered Trinity College in 1823 and made such an impression that in 1827 he was appointed Professor of Astronomy before even taking his degree. He is remembered for his prediction of conical refraction of light (confirmed experimentally by Humphrey Lloyd) but more particularly for his invention of quaternions, which had revolutionary and far-reaching consequences. The concept came to him in a flash of imagination as he was walking along the towpath of the Royal Canal where Brougham Bridge crosses it at Broombridge Road. Reputedly, he immediately scratched the crucial formula on the stonework of the bridge. If so, it has long since disappeared but today a memorial plaque commemorates the event. It reads:

$$i^2 = j^2 = k^2 = ijk = -1.$$

Hamilton believed that in quaternions he had found a natural algebra of three-dimensional space but this proved not to be the case. Sadly, he became pathologically depressed after two unhappy love affairs and a disastrous marriage: he sought refuge in alcohol, which clouded his later career.

The second Dublin-born genius was George Francis Fitzgerald (1851–1901), also a graduate of Trinity College. In 1881 he was appointed Professor of Natural and Experimental Philosophy there, a post he held until he died. He is particularly remembered for his hypothesis – the Fitzgerald Contraction – explaining the failure of the Michelson-Morley experiment (1880) on the velocity of light to detect the existence of an 'ether'. Put simply, this postulated that when bodies move through an electro-magnetic field they become very slightly contracted in the direction of motion. This hypothesis was independently developed by Hendrik Lorentz (1853–1928) in the Netherlands, and the effect is now commonly called the Lorentz-Fitzgerald Contraction. It was an important stage on the road to the theory of relativity propounded by Einstein. Fitzgerald lived at 19 Lower Mount Street: his portrait hangs in the Engineering School of Trinity College.

The geologist John Joly FRS (1857–1933) was born in Offaly and was educated at Trinity College, where he became Professor of Geology in 1897. He is remembered for an ingenious – but not very accurate – method of estimating the age of the Earth by measuring the salinity of the oceans. His assumption was that the amount of salt delivered to the oceans by rivers flowing into them had remained constant throughout geological time. This led him in 1898 to an age of 80 million years, later amended to 100 million. This was in remarkably close agreement with the age determined by Lord Kelvin (1824–1907) on quite different grounds. In the event, both were proved very wrong: the figure currently accepted is around 5000 million years. Joly also investigated the earth's heat content, and the contribution to it of the radioactive constituents of its crust.

The Rotunda Hospital – founded by Bartholomew Mosse in 1745 for maternity work among the poor – is a reminder of the status of Dublin as a medical centre. In this context, two names are internationally famous. Robert Graves (1796–1853) graduated in medicine in 1818 and was appointed physician to the Meath Hospital in 1821. He became Professor of Medicine at the Irish College of Physicians, of which he was President 1843/4. His *Clinical Lectures on the Practice of Medicine* (1848) earned him an international reputation but today he is remembered more for his identification of exophthalmic goitre (Graves' Disease) in 1835. His contemporary William Stokes (1804–78) was the son of the physician Whitley Stokes (1763–1845), who was appointed Regius Professor of Medicine at Trinity College in 1838. William first graduated in medicine in Edinburgh (1825) and later in Dublin, where he was appointed Regius Professor in succession to his father in 1845. With Graves, he reformed the teaching of medicine in Dublin. Today, his name is remembered in Cheyne-Stokes Respiration – in which there is a cyclic waning and waxing in the depth of respiration – and the Stokes-Adams Syndrome, marked by attacks of unconsciousness resulting from cerebral anoxia.

A major scientific event in the 20th century was the foundation of the Dublin Institute of Advanced Studies in 1940. Originally, it had two departments: for Celtic Studies and Theoretical Physics. A department of Cosmic Physics was added later. It was the brainchild of Eamon de Valera (1882–1975), famous in history as the politician largely responsible for the creation of the Republic of Ireland in 1922. Much less well-known is that he was by training a mathematician, graduating from the Royal University of Ireland in 1901. Briefly (1906) he was Professor of Mathematics at Carysfort Training College, Blackrock, but thereafter devoted himself entirely to Irish politics. Despite its hazards and demands he maintained an active interest in mathematics to the end of his long life. He was elected Fellow of the Royal Society in 1968. It is recorded that when he was over 80, and virtually blind, he would take walks, on his doctor's recommendation, only after his secretary had read him a mathematical problem which he could ponder as he went along. As Prime Minister he was conscious that, as a small and poor country, Ireland could achieve little in scientific fields requiring expensive equipment and large numbers of research workers. However, he identified theoretical physics as 'a branch of science in which you want no elaborate equipment, in which all you want is an adequate library, the brains and the men, and just paper . . . This is the country of Hamilton, a country of great mathematicians.' It was de Valera who organized the memorial to Hamilton on Brougham Bridge. The school got off to a flying start with the appointment as first Director of Erwin Schrödinger (1887–1961), founder of wave mechanics and Nobel Laureate in 1933. He had had to leave Austria at the time of Hitler's *Anschluss* but returned there, as Professor of Physics in Vienna, in 1957.

But Dublin can pride itself on a truly indigenous Nobel Laureate. This is Ernest Walton – born in Dungarvan, Co. Waterford in 1903 – who read physics and mathematics at Trinity College 1922–6. He then went to the Cavendish Laboratory, Cambridge, where he joined forces with John Cockcroft (1897–1967). Together, in 1932, they performed the historic 'atom-splitting' experiment, for which they were jointly awarded a Nobel Prize in 1951. In 1934 he returned to Trinity College, where he became Professor of Natural and Experimental Philosophy 1946–74. Since his retirement he has lived quietly in Dublin.

Although there have been plenty of small industries – such as mining, textiles, fishing and glass – Ireland has never been a notably industrial country and derived little benefit from the Industrial Revolution. Dublin itself is noted for only two major businesses. The first – known world-wide by its Irish harp logo – is Guinness' Brewery, founded by Benjamin Guinness in 1759: by 1900 it overshadowed all other local firms. The other is Jacob's, a well-known bakery founded about 1850: it became famous for its 'steam bakery and flour stores'.

A distinguished Dublin resident in the early years of this century was John Boyd Dunlop (1840–1921) who had been a very successful veterinary surgeon in BELFAST but is better known as the inventor (1888) and manufacturer of the pneumatic tyre. His small firm eventually grew into the giant Dunlop Tyre Company but Dunlop resigned his directorship in 1895. He moved to Dublin and contented himself with being chairman of Todd, Burn & Co., a large drapery.

For students of biology, as well as the general public, Dublin offers both a zoo and a botanic garden. The first, in Phoenix Park, was created by the Royal Zoological Society of Dublin in 1831, the year after it was founded, and has since been greatly expanded: it is famous for its success in breeding lions. The National Botanic Garden, at Glasnevin, two miles north of the centre of Dublin, was founded in 1795 by the Royal Dublin Society with the aid of a Parliamentary grant. Today, the Society is better known for staging the annual Horse Show, a major social event.

DUDLEY, W Mids

The Black Country Museum has a wide range of exhibits representing local industry – coal-mining, nailmaking, glass-cutting, chainmaking and iron working. The Broadfield House Glass Museum has a famous collection of glass, with emphasis on that from the STOURBRIDGE area.

DUFFTOWN, Gramp.

The Glenfiddich Distillery Museum has an organized exhibition depicting the history of distillation there since 1886, with coopers' tools, stills and equipment, and other memorabilia relating to the family of William Grant.

DUMBARTON, Strath.

The massive Dumbarton Rock has revealed little but is a site of historic interest: in the Dark Ages it was the capital of the British kingdom of Strathclyde, the name still given to this region of Scotland.

The Denny Tank, built in 1883 by William Denny III, was the world's first tank for testing the design of merchant ships. Over the years models of many famous vessels have been tested there, including the P.&O. liner *Canberra* and Thomas Lipton's *Shamrock III*.

DUMFRIES, D & G

A camera obscura was installed at the top of a converted windmill to observe Halley's Comet in 1835. Today it is used to provide panoramic views of the surrounding countryside. The Windmill serves also as a museum of local history, depicting many local trades and crafts.

DUNBAR, Loth.

Traprain Law (5 miles W) is one of the largest hillforts in Scotland, probably a tribal centre of the Votadini, but later occupied by the Romans. A fine collection of late Roman treasure discovered there is now in the Royal Museum of Scotland, EDINBURGH.

DUNDALK, Louth

The souterrain at Donaghmore (3 miles W) is one of the finest in Ireland. Five passages, with an overall length of nearly 100 m, have been laid out at various levels.

DUNDEE, Tays.

From 1822, when flax began to become scarce, the city became an important centre for processing jute, imported from India for the manufacture of carpets, sacks, linoleum, rope and many other products. The Central Museum has many exhibits related to the industry.

Technology, like other human activities, has its tragedies as well as its triumphs. One such was the rail bridge across the River Tay from Dundee to Fife. The original bridge, constructed by Thomas Bouch (1822–80), was completed in 1877, and for this and other achievements he was knighted in 1879. In the same year the bridge collapsed in a severe gale and a train plunged into the river with the loss of seventy-five lives. The subsequent inquiry showed that insufficient allowance had been made for wind forces. The shock was more than Bouch could bear and he died in the following year. The present bridge, some 2 miles long, is carried on seventy-three pairs of piers.

The Mills Observatory in Balgay Park houses a 10-inch refracting telescope available for public use. There is also an exhibition of telescopes and other scientific instruments. The *Discovery*, Scott's Antarctic research vessel, formerly moored on the Embankment in London, is now in the Victoria Dock (see CARDIFF).

Patrick Bell (1799–1869), inventor of one of the first successful mechanical reapers, was born at Auchterhouse (4 miles NW). It was conspicuously successful when exhibited at the Perth Show in 1852.

Sir Robert Watson-Watt (1892–1973), famous for his pioneer work on radar, was a graduate of University College, Dundee (now an independent university).

DUNFERMLINE, Fife

Dunfermline was once the world's largest producer of damask table linen and a representative collection is displayed in the Dunfermline District Museum (Viewfield Terrace). Also displayed is an 1835 Jacquard handloom and some specialist tools of the textile industry.

DUNGENESS, E Ssx

The two nuclear power stations dominate the flat coastline. Dungeness A is a Magnox-type station; Dungeness B incorporates an advanced gas-cooled reactor.

DUN LAOGHAIRE, Dub.

Birthplace of George Stoney (1826–1911), mathematical physicist. After graduating at Trinity College, DUBLIN, he worked for a time as astronomical assistant to the 3rd Earl of Rosse at BIRR. For five years he was Professor of Natural Philosophy at Queen's College, GALWAY, and then from 1857 he was appointed Secretary of Queen's University, DUBLIN.

DUNS, Bdrs

Until comparatively recently, tinplate boxes were widely used for packaging, especially in the food industry. The Biscuit Tin Museum (Manderston) displays some two hundred tins made for Huntley and Palmer since 1873.

DUNSFORD, Devon

Edge tools have been made for more than two centuries at Dunsford Edge Toolworks. The museum has a collection of such tools and demonstrates traditional working methods.

DUNSTABLE, Herts.

Whipsnade Zoo, an extension of the London Zoo in Regent's Park, has a wide-ranging collection of animals confined in large open-air enclosures.

DUNSTER, Som.

Dunster Castle was the home of the Luttrell family for six centuries, until acquired by the National Trust in 1976. An 18th-century water-mill in the grounds has been restored to working order.

DUXFORD, Cambs.

The famous wartime RAF station is now an outstation of the Imperial War Museum, London. It houses Britain's finest collection of civil and military aircraft, including Concorde.

DYFFRYN ARDUDWY, Gynd

A large chambered cairn contains two burial chambers.

DYLIFE, Powys

Today, the isolated Plynlimon village of Dylife consists of no more than a public house and a few scattered houses, but in the 1860s, its population exceeded a thousand and it was a thriving centre for mining lead, producing up to 2500 tons of rich ore annually. Huge profits were made and the owners included the social reformers Richard Cobden (1804–65) and John Bright (1811–89). Its prosperity depended partly on efficient machinery and equipment – trams loaded underground were hoisted to the surface from depths up to 300 m, a practice then almost unknown in metalliferous mines – and social concern: Dylife was the only such mine in Wales that provided changing accommodation for the workers. Extensive use was made of water power (and later steam) for pumping, hauling and ore dressing. One wheel was 63 ft in diameter and was the largest ever erected in Wales. Another was connected with its mineshaft by a drawing system involving a wire cable nearly a mile long: parts of the line of posts carrying the pulleys can still be seen. There is much surface evidence of all this activity, including the ruins of a chapel and a graveyard.

However, output later declined as the best ore was exhausted and the price of lead declined, a common fate of mines the world over. Nevertheless Cobden, Bright, and Co. managed to dispose of it for £73,000 in 1873. The last mining activity at Dylife was in the 1920s when an unsuccessful attempt was made to rework profitably the enormous spoil heaps resulting from earlier operations.

E

EAGLESFIELD, Cumb.

Eaglesfield, near Cockermouth, is the birthplace of John Dalton (1766–1844) who in 1803/4 conceived the fundamental idea that the atoms of different elements are distinguished by differences in their weights. Son of a poor weaver, he was educated at the local school and for a time taught there. While a pupil his interest in science was aroused by his schoolmaster Elihu Robinson, an experienced meteorologist and instrument maker. From early childhood until his death Dalton kept daily records of local weather. A memorial plaque on his old house reads: 'John Dalton, DCL, LLD – Discoverer of the Atomic Theory – was born here Sept. 5, 1766, died at Manchester July 27, 1844.'(See also KENDAL, MANCHESTER.)

Robert of Eglesfield founded Queen's College, Oxford, in 1340.

The area has strong Quaker connections. There is a Quaker graveyard at Pardshaw Hall in which Friends may be buried who die in places without their own cemetery. Appropriately, it includes a simple memorial to John Dalton: 'He was not for an age, but for all time.'

EARLS BARTON, Northants.

Two industries are particularly identified with Northamptonshire: leather and lace. Both are represented in the Earls Barton Museum (West St), where there is a colourful collection of bobbins made locally.

EASDALE ISLAND, Strath.

Easdale Island, 16 miles south-west of Oban, was an important centre of the slate industry from the 17th to the 19th century, with large exports to Australia and North America. In 1881 the quarries were flooded by a freak storm and the industry collapsed. Easdale Island Folk Museum has exhibits depicting the history of the industry and other aspects of local life.

EASTBOURNE, E Ssx

Birthplace of Sir Frederick Gowland Hopkins (1861–1947) – Nobel Laureate 1929, President of the Royal Society 1930–5 – noted for his development of the concept of vitamins. (See CAMBRIDGE.)

EAST BUDLEIGH, Devon

The James Countryside Collection illustrates the impact of mechanization on agricultural practice. It includes collections of agricultural machinery and implements and exhibits illustrating the life and work of rural craftworkers such as saddlers and thatchers.

EAST CARLTON, Leics.

The East Carlton Heritage Centre is devoted to the origins and development of the Northamptonshire iron industry, especially since the opening of the Lloyd's works in the 1880s. All aspects of steel-making are portrayed, from ore-mining to fabrication of the finished metal.

EAST DEREHAM, Nflk

Birthplace of W.H. Wollaston (1766–1828) whose technique of powder metallurgy rendered the intractable metal platinum malleable and brought it into general use. He also discovered the related metals rhodium and palladium. (See also LONDON.)

EAST GRINSTEAD, E Ssx

At Holtye (5 miles SE) a stretch of Roman road is preserved. The local iron industry provided furnace slag to make a firm surface. The ruts made by cartwheels centuries ago can still be seen.

EASTHAMPSTEAD, Berks.

Caesar's Camp is a particularly fine example of an Iron Age contour fort. The whole perimeter strictly follows the 122m contour, giving an outline shaped like an oak leaf.

EAST KILBRIDE, Strath.

Birthplace of the famous anatomists William (1718–83) and John Hunter (1728–93). Their home, Long Calderwood, was then a farm but still exists within a modern conurbation: the house bears a commemorative plaque.

EAST LINTON, Loth.

Preston Mill is the oldest working water-powered grain mill in Scotland. As with many Scottish mills designed to grind oats rather than wheat, there is an associated drying kiln.

EASTON, Sflk

Easton Park Farm displays a collection of agricultural machinery and farm equipment.

ECCLES, GM

Monks Hall Museum displays a collection of machinery associated with the engineer James Nasmyth (1808–90), remembered particularly for his steam hammer. In 1836 he leased 6 acres of land at Patricroft to build a factory which eventually became the famous Bridgewater Foundry.

EDINBURGH, Loth.

The capital city of Scotland, situated on the south side of the Firth of Forth: the Forth

Bridge is nearby at Queensferry. The University of Edinburgh was founded in 1583 and attracted scholars from all over Europe – now from all over the world. The extent and diversity of the city's intellectual life earned it the somewhat fulsome title of the Athens of the North: more aptly, until air pollution began to be taken seriously, it was known as Auld Reekie. Fellowship of the Royal Society of Edinburgh (22–24 George Street), founded in 1783, is still a prized distinction in academic circles. In 1966 the old Heriot-Watt Technical College was given an independent charter as Heriot-Watt University, with emphasis on applied science and engineering.

Edinburgh has never been a great industrial city but its special interest in printing and publishing deserves mention. William Smellie (1740–1795) had his Printing House – now marked with a memorial plaque – in Anchor Close. There he printed, and partly wrote, the first three-volume edition of the *Encyclopaedia Britannica* (1768), still the world's leading work of its kind, encompassing the whole span of human knowledge. Almost as well known is *Chambers Encyclopaedia*, first published by Sir William Chambers (1800–83) and his brother Robert in 520 weekly parts 1859–68. Chambers Street is named after him and his statue stands there. In the context of light industry Alexander Bain (1810–77) is noteworthy. He was associated with some important innovations in telegraphy, including the use of perforated paper strips to transmit messages. He also made some of the earliest electric clocks. In 1941 the Institute of Electrical Engineers and other bodies placed a memorial plaque on 21 Hanover Street, where he had a workshop 1844–8. Alexander Graham Bell (1847–1922), inventor of the telephone, is similarly remembered at 16 South Charlotte Street, where he lived with his parents before emigrating with them to North America in 1870. Bell's interest in sound derived from his father Alexander Melville Bell (1819–1905), who had a keen interest in speech education for the deaf and devised a system of 'visible speech'. Another man outstanding in the field of tele-communications was Allen Campbell Swinton (1863–1930), a versatile electrical engineer remembered as a pioneer of modern television. He made his career in London. Son of a Professor of Civil Law in the University, he was born and brought up at 9 Albyn Place: here, too, there is a plaque.

In February 1841, W.H. Fox Talbot (1800–77) patented his Calotype process and the age of photography dawned. The first Calotype portrait in Scotland – and possibly the first in the world – was taken in May of the same year by Dr Robert Adamson, who opened a portrait studio at Rock House, Calton Stairs. In collaboration with the Edinburgh painter David Octavius Hill (1802–70), he embarked on an immensely successful career as a fashionable portrait photographer. Among those who sat for him was the engineer James Nasmyth (1809–1890), son of the portrait painter Alexander Nasmyth (1758–1840), who ran an art school at his house at 47 York Place, where James was brought up. Hill's portraits are now keenly sought collectors' items.

Three other eminent engineers have Edinburgh associations. Thomas Telford (1757–1834), the great civil engineer, was born in Eskdalemuir and as a young man worked as a mason on the building of the New Town. Among his memorials are the Caledonian Canal, built 1804–22, and in Edinburgh Dean Bridge (1832). The Institute of Civil Engineers placed a plaque on the bridge in 1957 to mark the bicentenary of his birth. Robert Stevenson (1772–1850), lighthouse engineer, lived at 1 Baxter Place for the last twenty years of his life. Finally, the railway engineer Nigel Gresley (1876–1941), was born at 34 Dublin Street (then No. 14).

The roll-call of famous scientists with Edinburgh connotations is impressive. The most outstanding are James Clerk Maxwell (1831–79) and Sir Edward Appleton (1892–1965). Maxwell was born at 14 India Street, where there is a commemorative plaque, and was a student at the University: in 1871 he became the first Cavendish Professor of Experimental Physics at CAMBRIDGE. He is remembered as the formulator of the electromagnetic theory of light. Appleton was appointed Principal and Vice-Chancellor of the University in 1949. He is particularly remembered for his research on the upper atmosphere: a layer of ionized gas which reflects radio waves is named after him. He was awarded a Nobel Prize in 1947.

However, the story begins much earlier than either Maxwell or Appleton. John Napier (1550–1617) – 8th Laird of Merchiston, inventor of logarithms – was born and died in Merchiston Tower, Colinton Road. He also invented Napier's Bones, a simple system of numbered rods for effecting multiplication and division. The family vault is in St Giles Cathedral. Sir David Brewster FRS (1781–1868) was a graduate of the University and its Principal from 1859 until his death. He did original research in optics and his name is remembered in Brewster's Law governing the polarization of reflected light.

The Edinburgh Veterinary College – now the Faculty of Veterinary Medicine within the University – was founded by William Dick (1793–1866), son of a farrier in the days when farriers customarily treated sick horses. He was born at 27 Canongate and is buried in the New Calton Burying Ground. The site of the family forge in Clyde Street is now a bus station.

In the 19th century the two great advances in medicine were the introduction of ether anaesthesia in 1846 – which allowed surgeons to undertake much more radical operations – and antiseptic surgery, which greatly reduced post-operative sepsis. Sir James Young Simpson (1811–70), son of a baker in Bathgate, studied medicine in Edinburgh and was appointed Professor of Midwifery. In 1847 he discovered the anaesthetic properties of chloroform. From 1845 he lived at 52 Queen Street, where a stone plaque commemorates his achievement. Another plaque, on the North British Hotel, marks the site of the pharmacy of Duncan Flockart and Co. which prepared for him the chloroform he needed for his studies. Joseph Lister (Lord Lister 1827–1912), pioneer of antiseptic surgery, was Professor of Clinical Surgery in Edinburgh 1869–77, spending the rest of his career in LONDON. He lived at 9 Charlotte Street.

Still in the field of biological science, the zoologist Sir D'Arcy Wentworth Thompson (1860–1948) was born at 3 Brandon Street, Canonmills, and studied natural science at Edinburgh and Cambridge. He was President of the Royal Society of Edinburgh 1934–9. The marine biologist Sir Charles Wyville Thomson (1830–92) was also a student at Edinburgh and in 1869 was appointed Professor of Natural History. He made his name as scientific director on the circumnavigational voyage of the *Challenger* 1872–6. On his return he directed the Challenger Expedition Commission in Edinburgh, publishing fifty large volumes of reports on collections made during the voyage. Sir Arthur Conan Doyle (1859–1930) is generally remembered only as the creator of Sherlock Holmes, but he also had a not undistinguished medical career, including some useful excursions into forensic medicine. He was born at 11 Picardy Place, but this is now demolished and the commemorative plaque which once marked it is now in the courtyard of Huntly House Museum, Canongate; a substitute is now affixed to No. 2 (opposite). He later lived, from 1876 to 1880, at 23 George Street.

In 1670 the Edinburgh Physic Garden was founded at Holyrood by two physicians – Sir Robert Sibbald (1691–1722) and Sir Andrew Balfour (1630–94) – who were also founders of the Royal College of Physicians of Edinburgh (1681). There is a memorial to Sibbald in Greyfriars churchyard. The garden's Keeper was James Sutherland, first Professor of Botany in the University, who built up a fine collection from correspondents all over the world. From 1675 to 1763 it was located – as a notice indicates – at what is now the car park of Waverley Station. The present Royal Botanic Gardens, and the associated Arboretum, are situated at Inverleith. Edinburgh has been designated a World Heritage Site by UNESCO.

EGHAM, Sry

The grandiose Royal Holloway College, modelled on a Loire chateau, is a memorial to the philanthropist Thomas Holloway (1800–83), son of a Penzance innkeeper, who made his fortune by the sale of patent medicines. He founded it in 1876 to provide higher education for women; by the time it was formally opened by Queen Victoria in 1883 nearly £1 million had been spent on it.

EGLWYSWRW, Dyfed

Pentre Ifan is a fine example of a Bronze Age cromlech.

ELING, Hants.

Tide mills are not uncommon in Britain, but the Eling Tide Mill is probably the only one still operational in western Europe. One mill is working, and grinds flour; the other is the centrepiece of an exhibition depicting all aspects of mill operation at the beginning of this century.

ELLESMERE PORT, Ches.

Ellesmere Port is situated at the junction of the Shropshire Union Canal and the Manchester Ship Canal (see MANCHESTER) and consequently was an important transshipment area. Many aspects of this activity are depicted in the Boat Museum, which displays more than fifty craft, ranging from narrow boats to tugs. It is based on the old warehouses and other buildings of the Shropshire Union. Other exhibits include operational steam engines, boat building, and canal construction. A restored row of early Victorian cottages illustrate the domestic life of those who worked on the canals.

ELSECAR, S Yorks.

Elsecar Colliery provides an exceptional opportunity to see an 18th-century Newcomen engine, built in 1795 and in continuous use until 1923.

ELVINGTON, N Yorks.

The Bomber Command Station here was operational throughout the Second World War, and has now been restored to the state it was in then, including control tower, radio room and signals office. There are displays of wartime aircraft and also engines.

ELY, Cambs.

Stained glass windows, which enrich many great cathedrals and churches, represent a unique blend of art and technology. The Stained Glass Museum, in the Cathedral, is the only museum in Britain devoted to this highly specialized technique. The exhibits represent more than sixty windows from medieval to modern times, with associated explanations of technique and style. The 19th century is particularly well-represented.

EMBSAY, N Yorks.

The Yorkshire Dales Railway displays a large collection of historic industrial locomotives and rolling stock.

ENFIELD, Mddx

The government small arms factory at Enfield Lock was famous for the Lee Enfield rifle, for many years the standard weapon of the British Army.

Birthplace of Sir Joseph Bazalgette (1819–91), a civil engineer remembered for designing and constructing a complete new drainage system for the City of London 1855–89.

ENNISCORTHY, Wex.

The County Museum, located in a 13th-century castle, has a good general collection of local interest. There is also a section displaying a variety of old types of lighting devices.

ENNISKILLEN, Ferm.

Site of a well-preserved 12th-century round tower, part of a Celtic monastery that stood on an island in Lough Erne. Such towers served as both belfries and places of refuge.

EWELL, Sry

Wallpaper has long been a specialized product of the paper industry. The collection of wallpapers in the Bourne Hall Museum (Spring St), the earliest dating from the late 17th century, is one of the finest in existence.

EWENNY, S Glam.

Village famous for its pottery, still made there.

EXETER, Devon

Formerly an important centre for the woollen industry, with many fulling mills as evidenced by surviving warehouses on the quay. Exeter Maritime Museum displays one of the world's largest collections of boats, about 160 in all.

The fine astronomical clock in the Cathedral dates from the 16th century, replacing an earlier one installed about 1384. It is operated by a modern movement but the original is preserved in the Cathedral. The Cathedral Close is crossed by an elegant iron footbridge erected in 1814.

For a time, in the 1840s, it seemed that atmospheric traction might replace steam locomotives on the railways. In this system an iron tube with a continuous longitudinal slit along the top, sealed with a leather strip, was laid between the rails. A piston within the tube was connected by a bracket to the front vehicle of the train. When the pipe ahead of the train was exhausted by a stationary pump at the side of the track atmospheric pressure propelled it along. The system has been described as a rope railway with a rope of air. Brunel experimented with it on the South Devon Railway in 1844 but it proved a financial as well as a technological disaster. One of Brunel's pumping stations survives, as a listed building, at Starcross, with a small museum.

Exeter was the birthplace of Thomas Mudge (1717–94), highly skilled horologist and inventor of the lever escapement.

F

FAKENHAM, Nflk

Following the national conversion to North Sea gas in the late 1960s most of the old coal-gas plants – over a thousand in all – were demolished. The small plant at Fakenham, opened in 1825, is one of the few to have been preserved.

FAREHAM, Hants.

In 1792 Henry Cort (1740–1800), Navy Agent in London, took over Fareham Ironworks at Funtley and there in 1783 developed his puddling process for converting pig iron into wrought iron. It would in any event have been a major development in metallurgy but was particularly timely because Britain then imported two-thirds of her wrought iron from Sweden and Russia. During the Napoleonic Wars this source was severely restricted. Britain, during the first half of the 19th century, made more wrought iron than the rest of the world. This was partly due to the fact that Cort's patent was invalidated in 1790, leaving the field open to others.

FARNHAM, Sry

Birthplace of William Cobbett (1763–1835), the agricultural reformer remembered for his *Rural Rides*. The Farnham Museum (West Street) has an exhibition commemorating him and also one on the growing and processing of hops, an important local industry. Farnham is noted also for an early example of prefabrication in the building industry: the timbers of Westminster Hall were prefabricated here 1394–5 before being sent to London for assembly.

FAVERSHAM, Kent

The Fleur de Lis Heritage Centre is located in a 15th-century building and has a wide-ranging display of local industry which, perhaps unexpectedly, includes explosives. C.F. Schönbein (1799–1868) came from Italy to manufacture gun-cotton here in 1846 but a disastrous explosion the following year killed twenty-one men and the factory closed.

FAWSLEY, Northants.

Birthplace of the mathematician John Wilkins (1614–72). His Parliamentary sympathies – he married a sister of Oliver Cromwell in 1656 – led to his appointment as Warden of Wadham College, Oxford, in 1648. There he gathered around him a group of natural philosophers who ultimately (1660) formed the Royal Society of London. He was a versatile writer – ranging from a learned work on codes and ciphers to his *Discovery of a World in the Moon*, an early exercise in science fiction describing a journey to the Moon and its inhabitants.

FERNWORTHY, Devon

Here there is a stone circle of thirty blocks of local granite, bearing a remarkably close resemblance to one at Gors Fawr at NEWCASTLE EMLYN, Dyfed, on the other side of the Bristol Channel.

FETHARD, Tipp.

Fethard Folk Farm and Transport Museum (Cashel Rd) is housed in an old railway store-building dating from 1882. It displays a wide range of horse-drawn vehicles, agricultural machinery and bicycles. There is also a collection of domestic equipment.

FISHBOURNE, W Ssx

The Roman villa dates from the 1st century AD and was built in a lavish style. Unfortunately, the main residence lies under the A27 Chichester–Havant road but the northern wing has been preserved: some fine mosaics have been uncovered.

FLEETWOOD, Lancs.

A town at the mouth of the River Wyre founded in 1836 by Sir Peter Hesketh Fleetwood (1801–66) who had it laid out by the architect Decimus Burton (1800–81). His hope was that passengers to Scotland might go by train to Fleetwood and thence by sea to Glasgow, but this hope was dashed when the railway over Shap Fell was opened in 1847. The town's fortunes revived in the 1870s when it became important as a fishing port; this aspect of its life is depicted in the Museum of the Fishing Industry in Dock Street.

FLIXTON, Sflk

Norfolk and Suffolk Aviation Museum has exhibits particularly relating to East Anglia. Some sixteen aircraft are displayed in the open and the exhibition hall displays a range of aviation memorabilia going back to the First World War.

FOLKESTONE, Kent

The Cliff Railway on the Leas was built in 1885. With a gauge of 5 ft 10 in, it is hydraulically operated. Cherry Garden Upper Works is a former waterworks building, in which is displayed a considerable collection of steam plant, including an 1889 triple-expansion horizontal engine. The original workshop survives.

The English Terminal of the Channel Tunnel was opened in 1994.

Birthplace of William Harvey (1578–1657), discoverer of the circulation of the blood.

FORNCETT ST MARY, Nflk

The Forncett Industrial Steam Museum has acquired from various sources a score of stationary steam engines, some of which can from time to time be seen running. The oldest is a single-cylinder Corliss engine of 1873. The collection also includes some of the engines removed from Tower Bridge (LONDON) when this was converted to electric power.

FOREST OF DEAN, Glos.

Despite its present attractive woodland scenery, this area was at one time highly industrialized, especially for iron smelting and coal mining. Many surface remains of this activity are apparent but two centres are of particular interest.

Whitecliff Blast Furnace, just SW of Coleford – whose name derives from the local charcoal industry – was built in 1806: it was worked for only a few years and is well preserved.

The Old Ham Mine (now Clearwell Caves) about 5 miles S of Coleford was one of the sources of iron ore for Whitecliff. At Norchard, just N of Lydney, there is a steam centre displaying Great Western Railway locomotives.

In 1810 David Mushet (1772–1847) founded a steelworks at Coleford, later to become, under his son R.E. Mushet (1811–91), a world-famous centre for the manufacture of tough alloy steels for machine tools.

FOULBY, W Yorks.

Birthplace of John Harrison (1693–1776), horologist. In 1726 he invented the bimetallic grid-iron pendulum designed to neutralize the change in length due to changes in ambient temperature. His greatest achievement, however, was the marine chronometer – for which he eventually received a prize of £20,000 – which for the first time enabled navigators at sea to determine longitude precisely. (See LONDON.)

FOULRIDGE, Lancs.

The Foulridge Tunnel under the Pennines, on the Leeds and Liverpool Canal, was completed in 1796.

FOWEY, Corn.

Castle Dore (3 miles N) is a small Iron Age hillfort with literary connotations. Traditionally, it was the seat of King Marc, uncle of Tristan, betrothed husband of Iseult of Ireland. A 6th-century memorial stone reads (in Latin) 'Here lies Tristan, the son of Cunomorus'.

FRESHWATER, IoW

Birthplace of Robert Hooke (1635–1702), son of John Hooke, curate of All Saints Church. He lived there until his father died in 1648, when he went to Westminster School in London. The church stands at the top of Hooke Hill; at its foot is a memorial reading 'Robert Hooke, borne nearby 1635. Physicist, scientist, architect, and inventor'.

The Medina Camera Museum (Golden Hill Fort) has a considerable collection of still and ciné cameras. It has a particular association with Julia Cameron (1815–79), pioneer photographer who came to live here in 1860.

G

....

GALASHIELS, Bdrs

The Galashiels Museum is located within the old Peter Anderson woollen mill, in which a loom is driven by a restored water turbine. Associated exhibits depict the local importance of the woollen industry. In the modern mill all aspects of wool processing can be seen, from spinning to weaving.

GALSTON, Strath.

Robert Stirling (1790–1878) was minister here 1824–76 and is buried in the churchyard. He is remembered as inventor (1816) of the Stirling engine, an external combustion engine in which air (or a suitable gas) in a cylinder operates two pistons when external heat is applied. Its great advantage is that almost any fuel can be used. With his brother James, Stirling devoted much of his life to improving his engine but though a number were built they had limited success, as a high temperature is necessary to maximize efficiency.

In the 1930s, however, there was revived interest in the Netherlands and the USA. In 1985 the French submarine *Saga* was fitted with two Stirling engines, using helium under pressure as the working gas. A working model engine, presented by Stirling in 1827, is exhibited in the Department of Natural Philosophy, University of GLASGOW.

GALWAY, Gal.

Seaport at the head of Galway Bay. The University College was founded in 1908 as a constituent college of the National University of Ireland (see CORK, DUBLIN) but began as Queen's College (1845). Scientifically, it is noteworthy for George Stoney (1826–1911), Professor of Natural Philosophy 1852–7. He has a lasting claim to fame through coining the word 'electron' as the absolute unit of electricity. (See DUBLIN, DUN LAOGHAIRE.)

GATESHEAD, T & W

The Bowes Railway was one of several in the north-east designed to carry coal to rivers for onward shipping. It was built in ten sections, the oldest by George Stephenson (1781–1848) in 1826; seven were rope-hauled and three used locomotives. Part of the system was acquired by Tyne and Wear District Council in 1976, including some forty wagons, with many buildings and much equipment. The Bowes Railway Heritage Museum displays steam and diesel locomotives and is responsible for the world's only surviving rope-worked standard-gauge railway.

GIRVAN, Strath.

Girvan was once a centre for ring-net fishing and the McKechnie Institute (Dalrymple St) displays a collection of nets and ancillary equipment.

GISBURN, Lancs.

At Tom Varley's Museum of Steam the emphasis is on fairground steam engines made by the famous firm of Burrell and Aveling, but includes some other old traction engines and the earliest known Foster steam wagon.

GLASGOW, Strath.

Scotland's largest city and seaport, situated on the River Clyde. Though some sort of settlement may have existed here as early as the 5th century, Glasgow's rise to fame really dates from the Treaty of Union of 1707. This put Scottish ports on the same trading footing as English ones, and Glasgow was well placed to gain a substantial share of the growing Atlantic trade, dealing particularly in sugar and tobacco, in this respect challenging BRISTOL. Although now sadly in decline, shipbuilding grew to be an enormous industry in the 19th century. In 1812 the 30-ton *Comet* – plying the Clyde between Glasgow and Greenock – was the first paddle-steamer to run commercially in Europe. Her low-pressure engine is preserved in the Science Museum in London.

A key factor in the rise of the shipbuilding industry was, of course, the steam engine and it is thus appropriate that it was in Glasgow that James Watt (1736–1819) invented the separate condenser that greatly increased the efficiency of the Newcomen engine. As Mathematical Instrument Maker to the University of Glasgow he had been asked to repair a model Newcomen engine and it was in the course of this that the novel idea occurred to him. This historic model is now preserved in the University's Hunterian Museum. This was founded in 1807 to house the astronomical and other collections formed by the Scottish surgeon William Hunter (1728–93) in London. It has since been considerably extended and now includes the principal items in the Kelvin Collection of Scientific Instruments, of which the Department of Physics and Astronomy is the custodian. Among other noteworthy exhibits are a working model of Kelvin's harmonic analyser and a small Stirling air engine presented by its inventor, Robert Stirling (1790–1878) of GALSTON.

However, steam had come to Glasgow before shipbuilding. The textile industry – first linen and then cotton – was established in the second half of the 18th century, initially using water power. The first steam-powered cotton mill was built in 1792. A striking survival is the carpet factory of James Templeton & Co. overlooking Glasgow Green. Built in 1888 it was modelled – with the somewhat bizarre fancy of the day – on the Doge's Palace in Venice. A Glaswegian whose name is enshrined in the English language is Charles Macintosh (1766–1843) who in 1823 invented a waterproof rubberized fabric. Today, the word mackintosh (*sic*) is widely used to denote any sort of waterproof coat.

With the advent of the railways Glasgow became not only the focus of the Scottish system, with four mainline termini, but, from the 1830s, an important centre for building steam locomotives. The North British Locomotive Company, the result of a merger of existing firms in 1903, was then the largest manufacturer in Europe. These activities are well displayed by the Museum of Transport, which has a fine collection of Scottish-built locomotives, as well as tramcars, fire-engines and other vehicles.

In the textile industry an important process is bleaching the linen or cotton fabric to make it wholly white. In the 18th century this was commonly done by prolonged exposure to sunlight in bleach-fields, in which the Glasgow area abounded. At the end of the century Charles Tennant (1768–1838) revolutionized this section of the industry with the introduction of bleaching powder (chloride of lime). His St Rollox works, built in 1799 beside the Monkland Canal, grew to be in its time the largest chemical works in Europe.

Although manifestly a great commercial and industrial city, Glasgow is also a great cultural centre. Its University, founded in 1450, is the second oldest in Scotland after St Andrew's (1412), and among its academic staff have been numbered many famous scientists, engineers and physicians. In the 18th century mathematics flourished under Robert Simson (1687–1768) who was professor from 1711. On his death he left to the University his collection of mathematical books, then acclaimed as the most comprehensive in Britain: they are still preserved as the Simson Collection.

Glasgow also took the lead in the teaching of chemistry, originally allied with medicine. In 1747 the physician William Cullen (1710–90) was appointed to the first independent lectureship in chemistry in Britain. In 1751 he combined this with the professorship of medicine until he moved to EDINBURGH four years later to continue his academic career there. Among Cullen's most brilliant students at Glasgow was Joseph Black (1728–99) who himself moved to Edinburgh in 1751. There he achieved international fame by his researches on the chemistry of gases, the alkaline earths, and the nature of latent and specific heat.

Glasgow has a particular place in the history of fine printing, for it was there that Robert Foulis (1707–76) set up his business in 1741 and produced a series of magnificent editions of the classics, now much sought after by collectors. Two years later he was appointed Printer to the University and in 1752 the success of his books emboldened him to found an Academy of Fine Arts, anticipating by fifteen years the Royal Academy in London, and this was housed within the University. Teachers of engraving, painting and sculpture were brought from abroad but, sadly, the Academy did not survive Foulis's death.

In the 19th century Glasgow was associated with three outstanding leaders of science and medicine. The precocious William Thomson (Baron Kelvin of Largs, 1824–1906) matriculated in the University at the age of ten and after further study at CAMBRIDGE returned to Glasgow as Professor of Natural Philosophy 1846–99. He was among the scientific giants of the 19th century, making major contributions in the fields of electromagnetism and thermodynamics. His memoir *On the Dynamical Theory of Heat* (1851) is one of the truly great classics of science. From 1852 he collaborated with James Prescott Joule (1818–89) of MANCHESTER, famous for his precise measurements of the mechanical equivalent of heat. But Kelvin was not only a theoretician: he constructed a large number of instruments, including an improved mariner's compass, a siphon recorder for telegraphy and a tide predictor. In 1866 he was knighted for his contribution to the laying of the first Atlantic telegraph cable and he was made a peer in 1892. His old home at 11 Professor's Square still houses, in fine working condition, the extremely accurate double-pendulum astronomical clock that he built 1868–72. His statue stands in Kelvingrove Park, beside the River Kelvin.

Kelvin was the brightest star in a brilliant constellation: many of his contemporaries at Glasgow also achieved international fame. Joseph Lister (Baron Lister, 1827–1912) was Regius Professor of Surgery at Glasgow 1860–9 before moving to EDINBURGH and LONDON. He is universally recognized as a prime architect of modern surgery. The introduction of anaesthesia in the 1840s enormously increased the scope of surgeons but too often their patients died of post-operative infection. Lister's introduction of antiseptic surgery – inspired by the researches of Louis Pasteur in France – dramatically reduced in-hospital mortality. His 'antiseptic principle' was too revolutionary to gain immediate acceptance but he lived to receive virtually all the honours open to him. Like Kelvin, he served as President of the Royal Society; he was made a peer in 1897, and he was an original member of the Order of Merit (1902). Like Kelvin, his statue stands in Kelvingrove Park.

Another name internationally known in the history of science is that of William Jackson Hooker (1785–1865), Regius Professor of Botany at Glasgow 1820–41. During these years he made the neglected Botanic Garden into 'the equal of any in Europe', leading to his appointment as first Director of Kew Gardens, a post in which his son Sir Joseph Dalton Hooker (1817–1911) eventually succeeded him.

Also a contemporary of Kelvin at Glasgow was William Rankine (1820–72). After a successful career as a practising engineer he was elected Professor of Civil Engineering and Mechanics in 1855. He made notable contributions to the theory of heat engines and is also remembered – appropriately for a great shipbuilding centre – for important contributions to naval architecture.

Three other distinguished Glaswegians are James Young (1811–33), founder of the Scottish shale-oil industry; Alick Isaacs (1921–67), discoverer of interferon, a natural product having antiviral and antitumour activity; and Sir Monty Finniston (1912–91), metallurgist and industrialist, who after the Second World War was responsible for producing Britain's first plutonium.

Today, Glasgow can boast two universities. Strathclyde University was founded as part of the university expansion programme of the 1960s but it has in fact a history going back to the 1780s when John Anderson (1726–94), Professor of Natural Philosophy in the University, began a class in physics for working men. To ensure the perpetuation of his idea – for which most of his colleagues had no great enthusiasm – he bequeathed most of his estate for the foundation of an institute to promote the public good by the application of science. A few months after his death Anderson's Institution was formally incorporated. Over the years its basic purpose remained unchanged though it underwent several changes of name: from 1828–77 it was Anderson's University. It became affiliated with the older university in 1913 and eventually, in 1964, achieved full university status as the University of Strathclyde.

The new university elected as its first Chancellor a Glaswegian chemist. This was Alexander Todd (Lord Todd of Trumpington) a graduate of Glasgow University who was eventually appointed to the prestigious chair of organic chemistry at Cambridge. Like Kelvin and Lister before him he gained virtually every honour open to him including a Nobel Prize in 1957 and the Order of Merit in 1977. The choice appropriately acknowledged the institute's long record of achievement in chemistry. The first Professor of Chemistry (1830) was Thomas Graham (1805–69), who moved to University College, London, in 1837. He was the first President of the Chemical Society, founded in 1841, and in 1857 he was appointed Master of the Mint, an office once held by Newton. His statue stands in George Square. Before Graham's appointment, chemistry had been taught by George Birkbeck (1776–1841), among the few academics sympathetic to Anderson's ideas. In 1804 he moved to London to practise medicine, but remained enthusiastic about popular science teaching. In 1824 he founded the London Mechanics' Institute, which in the fullness of time developed into Birkbeck College, now part of London University.

GLASTONBURY, Som.

The 14th-century barn of Glastonbury Abbey is perhaps the most striking feature of the Somerset Rural Life Museum (Chilkwell St). Its roof timbers are an outstanding example of medieval wood working. It and the adjacent farm buildings depict many aspects of local rural life. An unusual exhibit is the mud-horse, a kind of sledge developed locally to enable fishermen to venture out on the mudflats of the Bristol Channel.

GLENCOLUMBKILLE, Dngl

The Folk Museum – housed in three cottages of the 18th, 19th and 20th centuries respectively – displays machinery and equipment used for the manufacture of Donegal tweed.

GLENCORSE, Loth.

Birthplace of C.T.R. Wilson (1869–1959), physicist famed for the invention (1911) of the Wilson Cloud Chamber for tracking the paths of single charged atomic particles. This immediately became an indispensable tool for atomic physicists world-wide. He was awarded a Nobel Prize in 1927.

GLENFIDDICH, Gramp.

The Glenfiddich Museum illustrates all aspects of whisky manufacture in Victorian times. The famous distillery was founded in 1886.

GLOSSOP, Derbs.

The Dinting Railway Centre (2 miles W off A57) displays a collection of historic steam locomotives and rolling stock.

GLOUCESTER, Glos.

The National Waterways Museum has a large collection depicting all aspects of the history of Britain's canals and waterways. In the dockland area, it is housed on three floors of an old warehouse. This itself is of historic interest, with massive wooden beams and cast-iron columns. The Robert Opie Museum of Packaging and Advertising is housed nearby in the Albert Warehouse.

The elegant bridge across the River Severn at Over was built in 1831 by Thomas Telford (1757–1834).

Birthplace, and home until his early twenties, of Sir Charles Wheatstone FRS (1802–75), pioneer of the electric telegraph.

GOATHLAND, N Yorks.

Roman engineers were famous as road builders, but while the lines of their roads criss-crossing the country are well-known, rather few structures survive: many are buried deep under modern roads or other constructions. Wade's Causeway is a section of a Roman road across the moor that is well preserved, though it has been partially restored.

GOLCAR, W Yorks.

The Colne Valley Museum comprises a group of three weavers' cottages displaying working handlooms, a spinning jenny and other weaving machinery. There is also a clogger's workshop, clogs formerly being the common form of footwear in the mills.

GOSPORT, Hants.

The Submarine Museum includes the Royal Navy's first submarine, *Holland I*, and HMS *Alliance* of the Second World War.

GOTT, Shet.

Tingwall Agricultural Museum depicts the life of the crofter/fishermen of Scotland. It is built on a working croft.

GRANGEMOUTH, Cent.

The Grangemouth Museum is largely concerned with the development of Sealock into the modern port of Grangemouth, with emphasis on waterways and shipping. Exhibits include a model of the *Charlotte Dundas*, the world's first practical steamship.

GRANGE-OVER-SANDS, Cumb.

The Lakeland Motor Museum, in the grounds of Holker Hall, houses a miscellaneous collection of motor-cars, motor-cycles and related memorabilia.

GRANTHAM, Lincs.

Isaac Newton (1642–1727) went to school here at the King's School, founded in 1328: there is a memorial plaque in the Old Schoolroom. As a boy Newton delighted to carve his name on every desk he occupied. None of these survive but his signature is on a stone window sill. His home was at Woolsthorpe Manor (5 miles SW). He returned there from Cambridge 1665–7 to escape the Great Plague and much of his most important work was done there, though not published until much later.

GRASSINGTON, N Yorks.

The 60-ft chimney standing in isolation on Grassington Moor is a reminder of the old method of lead smelting. The smelters themselves were located a mile away from the chimney, to which the lead-laden fumes were conveyed by 6-ft flues, from the inside of which lead was periodically scraped by hand. The Upper Wharfdale Folk Museum, housed in two old miners' cottages, includes exhibits relating to lead mining.

GREAT BEDWYN, Wilts.

Birthplace of Thomas Willis (1621–75), physician famous for his work on the anatomy of the brain. His *Cerebri anatome* is one of the great classics of medical literature. By tasting the sweetness of the urine in patients, he was virtually the first to recognize diabetes mellitus.

GREAT DRIFFIELD, Humb.

The great Iron Age cemetery at Danes Graves (3 miles N) once revealed some five hundred barrows, each 3–10 m in diameter but little over 1 m high. Of these only about two hundred now survive.

GREAT HARWOOD, Lancs.

Birthplace of John Mercer (1791–1866), pioneer of textile chemistry. He invented the process of 'mercerization' to make cotton fibres lustrous by treatment with caustic soda. His interest in calico printing led him to develop a chlorination process for wool to make it as easily printable as cotton: many years later it was discovered that this treatment also conferred valuable shrink-resistance properties. Mercer also invented photographic methods of printing on fabric. Queen Victoria was presented with a handkerchief so printed when she visited the Great Exhibition in 1851.

GREAT TORRINGTON, Devon

The crystal glassware produced by Dartington Glass is world-famous. The associated museum depicts the history of glassmaking since 1650.

GREAT WITCOMBE, Glos.

Site of a large Roman villa, enlarged in the 4th century. Substantial excavation has uncovered fine mosaic floors and a bath-house.

GREAT YARMOUTH, Nflk

Appropriately, the Maritime Museum for East Anglia is housed in what used to be a home for shipwrecked sailors. Devoted primarily to the local fishing industry and inland waterways, exhibits include a number of small marine engines. Among vessels preserved is the exotic lateen-rigged racer, *Maria*, built in 1827.

Burgh Castle (4 miles SW) is a Roman fort, part of the defence of the Saxon Shore. Much of the perimeter wall survives to a height of 5 m.

GREENFIELD, Clwyd

In the latter part of the 18th century the Greenfield Valley was a major industrial centre, the streams running fast off the mountains providing abundant water power. Copper derived from Parys Mountain, Anglesey (see AMLWCH) and smelted in

ST HELENS, was forged into rods and sheets. There were also two cotton mills. There are many apparent surface remains, to which the Greenfield Valley Heritage Trust provides a guide.

GREENOCK, Strath.

Birthplace of James Watt (1736–1819), inventor of the improved steam engine. The McLean Museum includes models of ships and engines and a few Watt memorabilia, including his patent letter-copying machine. A fine open-air statue of Watt stands near the Victoria Tower.

GREENSTEAD-JUXTA-ONGAR, Esx

The parish church is unique in being the only surviving example in Britain of a Saxon wooden building. It has interesting jointing. The split logs which form the walls are joined at top and bottom to wooden beams, using mortice joints. Each vertical is linked to its neighbour by tongue-and-groove joints.

GRESSENHALL, Nflk

As its name indicates, the Norfolk Rural Life Museum lays emphasis on local agriculture, of great importance in the economy of East Anglia for more than two centuries. It has an excellent collection of agricultural machinery and implements – much of it made by local engineering firms – and reconstructions of various craftsmen's workshops.

GRIMSBY, Humb.

Important centre of the fishing industry, especially after the introduction of steam trawling in 1881. The National Fishing Heritage Centre was voted the best industrial history museum in 1994.

GUERNSEY, Ch. Is.

Birthplace of Warren De La Rue (1815–89), pioneer of astronomical photography, especially of solar phenomena. He led the Spanish Eclipse Expedition in 1860 and demonstrated that the 'red flames' were solar, not lunar, in origin. His father was Thomas De La Rue (1793–1866), founder of the famous printing house with which Warren was actively associated throughout his working life.

La Houge Bie Museum (Grouville) is located by the gigantic mound of a Neolithic tomb, big enough to have had two medieval chapels built on it. Associated exhibits relate to the island's agricultural history and to the former Jersey Eastern and Western Railways, closed in the 1930s.

H

HADDINGTON, Loth.

Birthplace of John Rennie FRS (1761–1821), civil engineer responsible for a remarkable number of major projects, including three famous bridges in London (Waterloo, Southwark and the New London) and the Bell Rock lighthouse. He was totally dedicated to his work, making business appointments for 5 a.m.

HADRIAN'S WALL, Cumb., Northld

The four hundred years of Roman occupation of Britain left a lasting legacy – both cultural and physical – whose effects are still apparent. To effect the conquest begun by Claudius in AD 43, the Romans faced three major tasks. The first was to subjugate the new province in the military sense and this entailed the building of garrison forts, the remains of which abound, though often ill-preserved. Second, a system of roads had to be constructed so that troops and government officials could be moved quickly from place to place. Third, as subjugation proceeded, it was necessary to found towns as centres of regional administration: this was Britain's first taste of urban life.

In the event, the Romans never achieved total conquest. In the north, the fierce resistance, extended lines of communication and the difficult terrain combined to frustrate them. After less than a century, a halt was called on a line from Wallsend in Northumberland to Bowness-on-Solway, Cumberland. It was demarcated by the famous wall, commenced in AD 122 during the visit of Hadrian to Britain.

Hadrian's Wall is 75 miles long, punctuated by a small fort every Roman mile and every 5 miles by a larger fort with a garrison of up to a thousand men. Originally, parts of the wall were built in turf but by AD 136 much of this had been replaced by in-filled masonry, up to 3 m thick and 7 m high: this no doubt symbolized the Roman intention of permanent occupation.

Even today, with modern earth-moving equipment, the building of such a wall would be a considerable undertaking; in Roman times, with only manual labour and primitive tools, it represents an outstanding feat of civil engineering. Only local stone was used and a number of the grit-stone quarries have been identified; some are several miles from the wall itself. At the quarries the stones were pre-shaped to a taper, so that they would lock firmly together in position. From a tableau on Trajan's column in Rome (AD 110) it is clear that the stones were carried by the legionaries in baskets on their backs. The space between the two outer masonry walls was filled with rubble: possibly some of this was derived from the vallum (ditch) running parallel to the wall.

Hadrian's Wall was never intended as an extended defence line like, for example, the Maginot Line in recent times, but rather to make it possible to regulate traffic from the north: the 5-mile castles had gates permitting through traffic. In the event of attack troops could quickly be concentrated at the threatened point.

The first phase of Hadrian's Wall was brief. In AD 143 the frontier was pushed further north to the ANTONINE WALL and Hadrian's Wall was breached at many points, though the forts remained garrisoned. It was temporarily abandoned altogether in AD 167 but came into its own again about AD 200 when the Antonine Wall in turn was abandoned. Parts underwent several substantial rebuildings but Hadrian's Wall remained Rome's northern boundary until AD 383.

The wall is in various states of preservation. Its line can be seen throughout virtually its whole length but some of the more impressive sections are at Black Carts, Walltown Crag, Brunton and Poltross Burn (Northld). It has been designated a World Heritage Site by UNESCO.

HALESWORTH, Sflk

Birthplace of Sir Joseph Dalton Hooker (1817–1911), plant taxonomist and botanical explorer. Director of Kew Gardens 1865–85; President of the Royal Society 1872–7.

HALIFAX, W Yorks.

Noted medieval cloth-manufacturing town and still an important centre of the textile and other industries. Henry Lodge's cotton mill was the first factory in the world to be lit by gas.

The Calderdale Industrial Museum depicts almost every aspect of local industry, especially the woollen industry. It is housed in the magnificent Piece Hall where traders displayed their goods – a 'piece' being a length of cloth – which was opened in 1799. Exhibits include an original spinning jenny. Other industries represented are mining, quarrying, rope-making and leather.

Henry Briggs (1561–1631), who is remembered for bringing John Napier's discovery of logarithms to general notice, was born at Warley Wood. Halifax was also the birthplace of Jesse Ramsden FRS (1735–1800), famous maker of scientific instruments: he worked here as an engraver before setting up in business in London in 1755. One of his most important instruments was a 5-ft vertical circle he made for the observatory at Palermo.

Halifax is also the reputed birthplace of Sacrobosco of Holywood (d.1244 or 1256), one of the great medieval scholars and author of several books on mathematics and astronomy.

HALSTEAD, Esx

In 1828 Samuel Courtauld set up in business at Halstead as a silk-thrower and weaver: around the turn of the century the firm pioneered the manufacture and use of rayon, a semi-synthetic fibre made from cellulose, in COVENTRY.

The Brewery Chapel Museum depicts many aspects of local history but emphasizes the connection with Courtauld's Silk Mill.

HAMBLE, Hants.

The family grave of Sir Edwin Verdon Roe (1877–1958), aircraft manufacturer, is at

St Andrew's Church. He founded the Avro Aircraft Co. in 1910. The Avro 504 was one of the most famous military aircraft of the First World War: nearly ten thousand were built by his company and as many more by outside contractors. It remained in service for many years. Later, he turned his attention to flying boats.

HAMILTON, Strath.

The Hamilton District Museum (Muir St) is appropriately housed in an old coaching inn dating from 1696, for its considerable collection of horse-drawn and motor vehicles includes a four-in-hand coach.

HARDENHUISH, Wilts.

The large ornate monument by the SE corner of the church of St Nicholas is in memory of David Ricardo (1772–1823). Although remembered primarily as a political economist his early interest was in science. He set up a laboratory, formed a collection of minerals, and was a founder of the Geological Society (1807). In his *Principles of Political Economy and Taxation* (1817) he sought to apply scientific principles to economic problems.

HARLOW, Esx

The Mark Hall Cycle Museum has a very large collection of bicycles – from an 1818 hobby-horse – and accessories.

HARRAY, Ork.

The Corrigall Farm Museum, set in a restored 19th-century farmhouse, depicts rural life in Orkney in the 19th century and earlier. Interesting exhibits include a wooden plough and a wooden-wheeled ox-wagon.

HARTLEBURY, H & W

The Hereford and Worcester County Museum is housed in Hartlebury Castle, official residence of the Bishop of Worcester. The displays, mainly of local interest, cover various industries, including agriculture. They include a tailor's shop brought from Upton-on-Severn.

Sir Hiram Maxim (1840–1916) is best known for his machine-gun, adopted by the British Army in 1894. Apart from this, he was an extraordinarily versatile inventor – he called himself 'a chronic inventor' – whose interests ranged from aeroplanes and mouse-traps to merry-go-rounds. A Maxim Inhaler, to relieve bronchitis, among the exhibits, is a reminder of his versatility.

HARTLEPOOL, Dur.

Hartlepool has had a chequered history. In medieval times it was an important port but its fortunes declined and it became no more than a fishing village. In the 1830s came a revival, with the opening of the East Durham coalfield and the arrival of the railway. By the end of the century, it had become one of the biggest ports in the country, with a big shipbuilding industry. The history of these changes is depicted in the exhibits in the Hartlepool Maritime Museum (Northgate).

The prominent Nuclear Power Station – one of twelve operated by Nuclear Electric – incorporates an advanced gas-cooled reactor (AGR).

HASTINGS, E Ssx See WADHURST

HATFIELD, Herts.

Hatfield House has been the home of the Cecil family since 1611. There is an interesting botanic garden in the Knot Garden, in which are grown plants known in England in the 15th to 17th centuries. They include ones imported by John Tradescant (1570–1638), the head gardener, later famous for his own Botanic Garden in Lambeth, LONDON.

HAVEN STREET, IoW

The last steam passenger train on the Isle of Wight was acquired and maintained by a local consortium, which regularly runs services between Haven Street and Wootton. They also acquired the ancillary buildings – signal-box, workshops and other equipment.

HAWICK, Bdrs

In the 19th and early 20th centuries Hawick had an international reputation for hosiery and knitwear. This interest is well-represented – in terms of machinery and equipment – in the Hawick Museum at Wilton Lodge Park.

HAY-ON-WYE, H & W

At Dorstone (7 miles W) is Arthur's Stone, a Neolithic burial chamber. An enormous capstone, weighing about 25 tons, is supported by nine uprights.

HEBBURN, T & W

Birthplace of the geologist Arthur Holmes FRS (1890–1965), noted for his use of radioactive measurement techniques to determine the ages of rocks and thus the age of the Earth itself.

HECKINGTON, Lincs.

The tower windmill, built in 1830, situated just south of the village, is unique in being the only surviving eight-sailed windmill in Britain.

HELENSBURGH, Strath.

Birthplace of John Logie Baird (1888–1946), pioneer of television. (See LONDON.)

HELMSHORE, Lancs.

Two local museums illustrate many aspects – technological and social – of local industry. The Museum of the Lancashire Textile Industry, located in a mid-19th century mill, has an extensive range of machinery for processing and weaving cotton. A complementary collection is in the adjacent Higher Mill.

HELSTON, Corn.

At Wendron (2 miles N) is the Poldark Mine, now a tin-mining museum. It incorporates the former Holman Museum Collection. Major exhibits include two beam engines dating from 1830 and 1850 respectively. Some underground galleries are open.

At Cornwall Aero Park (by Culdrose Air Station) a collection of military aircraft is displayed in the open air. It includes helicopters and hovercraft.

HEMPSTEAD, Esx

In the 17th century Eliab Harvey, a wealthy London merchant, built himself a country residence here and had a family vault set aside in St Andrew's Church. Here many members of the family have been buried, most notably Eliab's brother Sir William Harvey (1578–1657), discoverer of the circulatory system of the blood. Apart from the vault, there is a marble bust and a large marble sarcophagus.

HENBURY, Avon

The Blaise Castle House Museum is basically a folk museum but among the exhibits is Stratford Mill. This water-mill was moved to the site from the Chew Valley when this was flooded to make a reservoir. A famous echo can be heard at Echo Gate.

HEPTONSTALL, W Yorks.

Heptonstall is one of many Pennine villages dependent over many years in the past on handloom weaving. This and other rural crafts are depicted in the Heptonstall Old Grammar School Museum.

HEREFORD, H & W

The Cider Museum displays cider-making equipment past and present, including a still for making cider brandy. In the Cathedral is one of the oldest surviving maps of the world, the famous 14th-century Mappa Mundi.

HERRINGFLEET, Sflk

Herringfleet Mill, a smock mill, is a survivor of the many windmills used in East Anglia for land drainage. In this instance, drainage was effected not by pumping but by a scoop-wheel.

HESSLE, Humb.

Whiting – washed and ground chalk – has long been used in whitewash and as a mild abrasive. Hessle Whiting Mill, built about 1810, is unusual in having five sails; it is no longer operational but most of the machinery survives. The mill not only drove rollers to grind chalk from an adjacent quarry, but also two pumps, one to draw water from a well and the other to pump slurry to settling tanks.

HEXHAM, Northld

Vindolanda (3 miles W), an important Roman camp just south of HADRIAN'S WALL, has been excavated, uncovering the 4th-century fort and part of the external town. Material recovered is displayed in the adjacent Chesterholm Museum. The examples of textiles and leather are particularly interesting.

The Roman fort at Chesters (5 miles N) has much of archaeological interest, including the exposed foundations of the barracks and the headquarters building (*principia*). Downstream is the military bath-house, with lockers for clothing surviving. At low water the piers of the Roman bridge over the Tyne can be seen. In the same area is the Roman fortress at Homesteads (Vercovicium). Among less usual items excavated, in good order, is the public latrine.

HEYSHAM, Lancs.

The two nuclear power stations dominate the coast south of Morecambe. Each incorporates two advanced gas-cooled reactors (AGRs), enclosed in pressure vessels made of pre-stressed concrete lined with steel.

HIGH WYCOMBE, Bucks.

The making of furniture has long been an important industry in the Chilterns, particularly rich in beech trees. The Wycombe Chair Museum depicts the equipment and techniques used in the industry. Towards the end of the 17th century Buckinghamshire became also a centre for lace-making and this is represented in a separate exhibit.

The town is the birthplace of Sir Geoffrey de Havilland (1882–1965), aeronautical engineer and industrialist.

HINKLEY, Som.

The two nuclear power stations at Hinkley Point dominate the flat local landscape. Hinkley A incorporates a Magnox reactor; the more recent Hinkley B an advanced gas-cooled reactor (AGR).

HITCHIN, Herts.

Birthplace of Sir Henry Bessemer (1813–98), steel manufacturer who invented the steel-making process that bears his name.

HOLYHEAD, Anglesey

Three Roman hillforts were located at Holyhead as protection against Irish pirates. Caer Gybi overlooks the harbour and parts of the walls, including corner towers, survive up to a height of 5 m. The others are Caer Twr and Caer Ty Mawr.

HOLYWELL, Clwyd

Grange Cavern Military Museum, in a former limestone quarry, houses over seventy military vehicles, both tracked and wheeled. Exhibits include a Chevrolet – probably the only one surviving – used by the Long Range Desert Group in N. Africa.

HONITON, Devon

Honiton is a famous centre for the hand lace-making industry. This is featured in the Allhallows Museum, located in a medieval chapel.

HORNSEA, Humb.

As its name indicates, the Hornsea Museum of Village Life mainly portrays local history. Two exhibits are of industrial interest: the Hornsea Brick and Tile Works (1868–94) and the Hull and Hornsea Railway (1864–1964).

HORSEY, Nflk

Horsey Mere Wind Pump, a tower mill, was built as recently as 1912 – but on the site of an 18th-century building – for land drainage.

HOVE, E Ssx

The British Engineerium is housed within the old Goldstone Pumping Station. A collection of large steam engines and many models illustrate the history of steam power.

HOWTH, Dub.

The National Transport Museum Project in Howth Castle has accumulated more than a hundred vehicles of all kinds, ranging from trams to buses, horse-drawn vehicles to fire-engines.

HOY, Ork.

The most famous landmark on Orkney is the Old Man of Hoy, a towering rock pillar. Hoy is also noteworthy for the Dwarfie Stane, a tomb hollowed out within a solid block of sandstone and approached by a passage 3 m long. Such tombs are common on the Iberian Peninsula, but this is the only example known in Britain.

HUDDERSFIELD, W Yorks.

Town long famous for its woollen goods, initially established as a cottage industry. In the mid-18th century, small water-powered mills grew up along the River Colne. The Tolson Museum, Ravensknowle, has exhibits illustrating the history of local industry.

HULL (KINGSTON-UPON-HULL), Humb.

The Humber Bridge, completed in 1981, is the world's largest single-span suspension bridge. Of the city's several museums three are of particular technological interest: the Town Docks Museum, which has a collection of scientific instruments, including Captain Cook's thermometer; the Hull Museum of Transport; and the Yorkshire Water Museum (part of an operational pumping station).

Birthplace of Adrian Haworth (1768–1833), naturalist. He spent much of his life at Cottingham and helped to found and arrange the Hull Botanic Garden. His *Lepidoptera Britannica* (1803–28) was for 50 years the standard work on British butterflies. Burton Constable Hall (at Sproatley 10 miles NE) was the home of William

Constable FRS (1721–91), botanist. The remains of an important collection of scientific instruments are housed there.

HUNGERFORD, Wilts.

At Littlecote Park (2 miles N) the foundations of a 4th-century Roman villa have been uncovered, together with a well-preserved mosaic floor.

HUNNINGTON, W Mids

Blue Bird has been an important part of the confectionery industry since 1895, when the firm was founded in Birmingham; in 1926 it moved to Hunnington, where the firm has established a museum. This features a wide range of tins (cf the Robert Opie Museum, GLOUCESTER) and other memorabilia.

HUNSBURY, Northants.

Northamptonshire has long been an important source of ironstone, supplying the local ironworks. The Northampton Ironstone Railway Museum is located on the railway serving the Hunsbury Hill quarries, now closed. Exhibits include locomotives, rolling stock and equipment used in the iron industry.

HUTTON-LE-HOLE, N Yorks.

The Ryedale Folk Museum has a good general collection, including 19th-century tools, but is particularly noteworthy for its unique Elizabethan glass furnace. It was originally built out on the moors at Rosedale, but has been dismantled and reassembled here.

IJ

IDBURY, Oxon.

Sir Benjamin Baker (1840–1907), remembered for many major civil engineering works but particularly for building the Forth Bridge (1890), is buried at St Mary's Church. There is a large memorial stone in the graveyard.

ILCHESTER, Som.

Birthplace of Roger Bacon (1214–1292). Medieval scholar – the *Doctor mirabilis* – commonly regarded as the first scientist in the modern sense. Much of his work was done in OXFORD.

ILKLEY, W Yorks.

Ilkley Moor is rich in Stone Age remains, including circles and rock carvings. The most famous is the Swastika Stone, possibly of Iron Age date, with a double swastika carved on it.

IMMINGHAM, Humb.

Immingham is an important port on the River Humber: the docks were built in 1912. Immingham Museum and Art Gallery depicts local history and includes a reconstruction of a chemist's shop as it was 1912–40.

INVERARAY, Strath.

Bell-founding was an important minor industry from medieval times: the wealth of Bell Fields around the country is a reminder that the casting was often done on the spot. Inveraray Bell Tower is noteworthy for containing the world's third heaviest ring of ten bells.

INVERNESS, Hghld

Birthplace of Sir James Swinburne (1858–1958), electrical engineer who made important contributions to the generation and distribution of electricity and to the improvement of the incandescent filament lamp. He also developed phenol formaldehyde resin as an industrial plastic.

Clava Cairns lie about 1 mile SE of the battlefield of Culloden. They comprise an interesting group of passage-graves and cairns.

INVERURIE, Gramp.

Balquhain, a small prehistoric stone circle, is situated 3 miles NW. Near by, at Loanhead of Daviot, is another.

IPSWICH, Sflk

East Anglia – a rich farming area – has long been a centre for the manufacture of agricultural machinery. In 1832 Ransomes – an engineering firm founded by Robert Ransome (1795–1864), a local iron founder – acquired a licence to manufacture the mechanical lawn mower invented in 1830 by E.B. Budding of STROUD: they are now the biggest manufacturers in Europe.

The general collection of the Ipswich Museum includes a number of scientific instruments. It also contains replicas of the Sutton Hoo and Mildenhall treasures (originals in the British Museum, LONDON).

Sir George Airy (1801–92), Astronomer Royal 1835–87, is buried in St Mary's Churchyard at Playford (5 miles NE). His memorial is a frontal bust set in an oval medallion.

IRONBRIDGE, Shrops. See TELFORD

IRVINE, Strath.

The Scottish Maritime Museum displays large and small ships and boats, boatbuilding methods, a shipyard worker's tenement flat and many items of maritime interest.

JODRELL BANK, Ches.

The world's first giant radio telescope is a landmark for miles around. Designed by Sir Bernard Lovell, of the University of MANCHESTER, it has a 250-ft steerable disk and was completed in 1957. As well as recording radio waves from space it has been used for tracking satellites and space probes.

JOHNSTOWN CASTLE, Wex.

The Irish Agricultural Museum illustrates the history of agriculture and rural life in Ireland. It is situated in early 19th-century farm buildings, the main features of which have been retained.

K

KEALKIL, Cork

Site of a small stone circle with five stones still upright: also two standing stones adjacent.

KELSO, Bdrs

The small Kelso Museum was once a skinner's workshop, and has a variety of exhibits illustrating this trade.

Birthplace of Sir William Fairbairn (1789–1874), engineer, who claimed to have built nearly a thousand bridges in his lifetime. His most impressive memorial is the tubular bridge over the Menai Straits, built with Robert Stephenson (1803–59). (See BANGOR.)

KENDAL, Cumb.

A town once famous for the manufacture of snuff, now no longer fashionable. John Dalton (1766–1844) was principal of a Quaker school here 1785–93. He, and the geologist Adam Sedgwick (1785–1873), were members of the Kendal Literary and Scientific Society, whose collection was transferred to the Kendall Museum in 1910.

The town has very strong Quaker connections. The Meeting House (Stramongate) displays a modern tapestry depicting the history of Quakers in the UK. One panel is devoted to scientists, including Dalton and Sir Arthur Eddington (1882–1944), pioneer of stellar structure and of the General Theory of Relativity, who was born here.

KESWICK, Cumb.

Home of Jonathan Otley (1766–1856), geologist and instrument maker: a few of his instruments are housed in the Keswick Museum and Art Gallery.

Graphite, or black lead, is mined locally (see SEATHWAITE). The Cumberland Pencil Museum illustrates methods of pencil manufacture from the 16th century. Keswick Railway Museum (Main Street) houses a small collection of Cumbrian railway memorabilia.

The great stone circle of Castlerigg, one of the finest in Britain, lies 1½ miles E.

KETTERING, Northants.

The NORTHAMPTON area is an important centre for the leather industry. The Westfield Museum here has a section devoted to the manufacture of boots and shoes. (See also STREET.)

KIDWELLY, Dyfed

The Kidwelly Industrial Museum is housed in the remains of a tinplate works. It includes much machinery, especially related to metal working and coal mining.

KILBARCHAN, Strath.

The cruck-built Weaver's Cottage was built in 1723 and used for weaving up to the Second World War. It has been restored to its 19th-century condition, and exhibits various machines and tools of the weaving industry. The weaver's shop in the basement has been completely restored and includes the loom and accessories used by a famous local weaver, Willie Meikle.

KILLARNEY, Kerry

The Transport Museum of Ireland displays a collection of cars dated 1901–22, bicycles and tricycles, and motoring memorabilia.

KILLHOPE, Dur. See COWSHILL

KILLINGWORTH, T & W

The engineer George Stephenson (1781–1848) lived from 1805 to 1823 in Dial Cottage (Great Lime Road), which carries a memorial plaque.

KILMARTIN, Strath.

The Kilmartin Valley is rich in prehistoric remains – circles, hill forts, standing stones, cup-and-ring marked stones, and cairns. Of particular interest is Temple Wood stone circle and Nether Largie North cairn with its stone carvings.

KILMODAN, Bdrs

Birthplace of Colin Maclaurin (1698–1746), mathematician. He was a strong advocate of Newton's new ideas and developed his work on geometry, fluxions and gravitation. (See EDINBURGH.)

KINCARDINE-ON-FORTH, Fife

Birthplace of James Dewar (1842–1923), professor in the Royal Institution, LONDON, 1877–1923. The house, now the Unicorn Hotel, bears a memorial plaque. He is famous for his research in low-temperature physics and as inventor of the now-ubiquitous vacuum flask.

KINGSBRIDGE, Devon

Birthplace of William Cookworthy (1705–80), a Plymouth pharmacist who discovered china clay in Cornwall and experimented with its use to make a true porcelain; he patented a manufacturing process in 1768. He began manufacture at Coxside, Plymouth, in 1770 but two years later moved to BRISTOL. The Cookworthy Museum commemorates this important invention, and displays many aspects of local crafts.

KINGSWINFORD, W Mids. See STOURBRIDGE

KIRKCALDY, Fife

Linoleum – so-called because, like paint, it is based on linseed oil – was first manufactured in 1864, in Staines, Surrey; it remained popular until the 1950s when it began to be replaced by various kinds of plastic sheeting. Latterly, it has begun to come back into fashion. The industry soon came to be identified with Scotland and the Kirkcaldy Museum has a special section devoted to it, among other local industries.

KIRKDALE, N Yorks.

Over the south door of the church of St Gregory is a Saxon sundial, commemorating Edward the King (presumably Edward the Martyr) and Earl Tosti.

KIRKSTALL, W Yorks.

The Kirkstall Museum of Social History was formerly the Great Gate House of the 12th-century Kirkstall Abbey. It features shops and workshops representing various trades, including that of saddler, nailer, clock and watchmaker, weaver and blacksmith. Much of the material was acquired from buildings in Leeds scheduled for demolition.

KIRRIEMUIR, Tays.

The geologist Charles Lyell (1797–1875) was born here on the Kinnordy estate. Although he himself never resided here he made several visits in later life to make local geological field trips.

KIRTON IN LINDSEY, Lincs.

The Mount Pleasant Windmill, a tower mill built in 1875, is still operational, though power is now supplied by a vintage steam engine. The adjacent Lincolnshire and Humberside Railway Museum represents most aspects of local railways, including a steam locomotive.

KYLE OF LOCHALSH, Hghld

Two large, well-preserved brochs – Dun Telve and Dun Troddan – are located at Glenelg (7 miles SE) (cf STORNOWAY).

L

LACOCK, Wilts.

W.H. Fox Talbot (1800–77), inventor of the crucially important negative/positive process of photography, lived in Lacock Abbey. His earliest surviving paper negative – showing a lattice window in the Abbey (1835) – is now one of the treasures of the Science Museum, LONDON. The associated Fox Talbot Museum illustrates the history of photography. His grave is in Lacock Cemetery. In the town is a large tithe barn with fine roof timbers.

The Lackham Agricultural Museum (Lackham College of Agriculture), located in reconstructed farm buildings, displays a range of agricultural tools and machinery.

LAMPETER, Dyfed

The Romans mined for gold at Dolaucothi (7 miles SE) and traces of their extensive workings can still be seen; a fort was built about AD 75 to protect them. There is evidence that they used the process known as 'hushing' to wash away the topsoil and expose the gold-bearing veins of ore. Some recovered artefacts can be seen in the CARMARTHEN Museum. Mining on a modern commercial basis was commenced in 1888 and continued until 1938, though with no great success.

LANCASTER, Lancs.

Small manufacturing city whose charter dates from 1193. The University was founded in 1964. It is the birthplace of three distinguished scientists/technologists. The ironmaster Henry Cort (1740–1800), known as the 'Great Finer', is remembered as the inventor of the puddling process for converting crude pig iron to wrought iron. He also introduced grooved rollers for working iron in place of the laborious hand hammering. (See also FAREHAM.) William Whewell (1794–1866), son of a joiner, rose to be a Fellow of the Royal Society, Master of Trinity College, Cambridge and Vice-Chancellor of Cambridge University. His knowledge was encyclopaedic: in his *History of the Inductive Sciences* he coined the word 'scientist' in its modern sense. Sir Ambrose Fleming (1849–1945), inventor of the thermionic radio valve, was also a native of Lancaster.

The Lancaster Maritime Museum is housed in the old Custom House (1764) and depicts the life of Lancaster when it was a busy port – trading particularly with the West Indies – in the 18th and early 19th centuries. The displays emphasize the importance of shipbuilding and related activities: there are a number of ship models.

LANEAST, Corn.

Birthplace of John Couch Adams (1819–92), eminent astronomer, discoverer of the planet Neptune. He was born at Lidcot Farm: a bust is displayed in Lawrence House,

Launceston, a local museum (NT). He was appointed Professor of Astronomy at CAMBRIDGE in 1858. (See also TRURO.)

LANERCOST, Cumb.

Thomas Addison (1793–1860), famous physician, is remembered particularly for his identification of the cause of Addison's disease, which results from adrenal insufficiency. He is buried at Lanercost Priory.

LANGTON MATRAVERS, Dors.

The quarrying and dressing of stone has been an important local industry since Roman times, and the Coach House Museum illustrates all aspects of the quarrying industry.

LANREATH, Corn.

The Lanreath Farm and Folk Museum is devoted to local agricultural history. It includes some unexpected exhibits: an 1865 lawnmower (see STROUD), an ice-cooled refrigerator, and a whimble for making rope from straw.

LAXEY, IoM

The huge Lady Isabella waterwheel (22 m diam.) – the largest in the world – was built in 1854 to drain local zinc mines. It developed 200 HP and could pump 250 gallons of water a minute from a depth of 1200 ft. It remains as a spectacular tourist attraction. The two main mines were Laxey and Foxdale, on the other side of the island, producing mostly lead and zinc.

LEEDS, W Yorks.

The city originated in Roman times at a ford over the River Aire and became an important centre for cloth manufacture in the 18th century. The fast-running tributaries of the river provided soft water and a source of power before the introduction of steam. The 127-mile Leeds and Liverpool Canal received Parliamentary approval in 1768 but was not completed until 1816. Major features are the famous five-rise staircase lock at BINGLEY (1774); the FOULRIDGE Tunnel (1796) through the Pennines; the embankment across the Calder Valley at BURNLEY (1799), and the 23-lock flight at WIGAN. The Canal Warehouse, Canal Basin, is a superb building dating from 1773. Other waterside warehouses are to be found in The Cells, a street dating from medieval times. Atkinson Grimshaw, the landscape painter remembered for his evocative pictures of Victorian Leeds, lived at 56 Cliff Road from 1866 to 1870. The Armley Mills Museum is located in one of Britain's largest woollen mills, built in 1806 and working until 1969. It illustrates every aspect of local industrial history and includes much machinery and engines, including a water-wheel. Later, Leeds became the 'city of a hundred trades' with textiles represented increasingly by ready-made clothing. The main diversification was into engineering for industrial and agricultural machinery. The Royal Armouries Museum, part of the National Collection of Arms and Armour at the Tower of London, is the main national museum for the history of arms and armour.

In 1758 the Middleton Colliery was joined with the Aire and Calder Navigation by a tramway. In 1812 steam traction was introduced using a rack-and-pinion system – designed by Matthew Murray (1765–1826), a local engineer – in which a toothed cog on the locomotive engaged with a toothed rail. It was thus the world's first commercial railway but with the availability of improved rails the system was soon abandoned. A model of Murray's locomotive can be seen in the Science Museum, London. The city retains some bizarre examples of industrial architecture. John Marshall's flax mill (1838–40) in Marshall Street was designed by Joseph Bonomi II (1796–1888) in the then very fashionable Egyptian style. T.W. Harding's pin works has a ventilation tower based on Giotto's campanile in the Duomo in Florence and a chimney modelled on the tower of the Palazzo de Signore, Verona. Matthew Murray and Sir Peter Fairbairn (1799–1861) made major improvements to textile processing machinery. Fairbairn House, 71 Clarendon Road, was built in 1840.

John Smeaton FRS (1724–92), famous engineer and instrument maker, was born at Austhorpe and educated at Leeds Grammar School. He made his name in London but returned to Austhorpe in 1756. There he built adjacent to his house a four-storey tower fitted up as a workshop and study. He is remembered particularly for his work on the improvement of wind and water mills and for rebuilding the Eddystone Lighthouse (see PLYMOUTH). The chemist Joseph Priestley (1733–1804), discoverer of oxygen, was born at Birstal Fieldhead (5 miles SW) and in 1767 became minister at Mill Hill Chapel (since rebuilt) in City Square, where there is a memorial statue. A more recent chemist associated with the city was Samuel Sugden FRS (1892–1950), who was born there and educated at Batley Grammar School. He is remembered for his concept of the 'parachor', relating molecular weight and surface tension. Leeds was for a time the home of Samuel Smiles (1812–1904), the social reformer who was editor of the *Leeds Times* 1838–42. For 21 years he was actively engaged in railway projects – notably the Leeds and Thirsk Railway – but increasingly gave his time to writing, extolling the virtues of Victorian enterprise. He is remembered for such books as *Self-Help* (1880), *Character* (1871), *Thrift* (1875) and *Duty* (1880). He also wrote biographies of several great industrialists, including George Stephenson, Josiah Wedgwood, and Boulton and Watt. One of his sons William Holmes Smiles (1846–1904) spent most of his working life in BELFAST, where in 1876 he established a small ropeworks, which ultimately became very large and prosperous. In the 1880s Louis Le Prince was a pioneer of cinematography, introducing a camera with a rotating battery of lenses. Appropriately, the site of his workshop in Woodhouse Lane is now occupied by the local BBC studios.

The University derives from the Medical School (1831) and the Yorkshire College of Science (1874). The two combined in 1884 to form (with LIVERPOOL and MANCHESTER) a constituent college of the Victoria University. It was granted its charter as an independent university in 1904. In his later years Leeds was the home of W.T. Astbury FRS (1898–1961) who in 1928 was appointed Lecturer (later Reader) in Textile Physics at Leeds University. He did pioneer research 1937–61 on the X-ray analysis of the structure of polymers, especially fibres. The city also boasts two Nobel Laureates, A.J.P. Martin (b.1910) and R.L.M. Synge (1914–94), who worked for many years in the laboratories of the Wool Industries Research Association. During the Second World War they developed an important technique of chemical analysis known as partition chromatography.

St James's is Europe's biggest teaching hospital. With it is associated a large medical museum housed in the old workhouse from which the hospital evolved. Exhibits include a large collection of instruments made by the local firm of Chas. F. Thackray Ltd.

LEEK, Staffs.

In 1742 James Brindley (1716–72), who had been an apprentice wheelwright in MACCLESFIELD, settled in business in Leek and began to work on the flint mills needed by Josiah Wedgwood (1730–95) for his pottery business. Although never more than semi-literate, he went on to make a great name for himself as a canal builder, sometimes in association with the Duke of Bridgewater. The James Brindley Museum displays many items relating to his life and work: it is housed in the restored Brindley Water Mill. At Cheddleton (5 miles S on A520) is an operational water-powered flint mill dating from about 1760.

LEGANANNY, Ant.

Here is to be seen Ulster's best-known dolmen – a huge capstone supported by three uprights.

LEICESTER, Leics.

An important centre of the knitwear and hosiery industry. The Leicester Museum of Technology displays giant beam engines, a steam shovel and other machinery, including some used for knitting. The John Doran Gas Museum is devoted to the history of the gas industry, especially in the East Midlands.

Birthplace of H.W. Bates FRS (1825–92), naturalist and Amazonian traveller, who discovered the phenomenon of mimicry in animal coloration. His *Naturalist on the River Amazon* (1863) is widely regarded as one of the best travel books ever written.

LEIGHLINBRIDGE, Carl.

Birthplace of John Tyndall (1820–93). After working for a time in the Ordnance Survey of Ireland he studied chemistry under R.W. Bunsen (1811–99) in Marburg. In 1843 he was appointed Professor of Natural Philosophy in the Royal Institution, LONDON, and in 1867 succeeded Faraday as Superintendent there.

LEISTON, Sflk

East Anglia is a prosperous farming area and has long been an important centre for the manufacture of agricultural machinery. Garrett's was established as a blacksmithing business in 1778 and its famous Long Shop – now a museum – was built in 1853 to assemble steam traction engines.

Close by, on the coast at Sizewell, are the massive buildings of two nuclear power stations. Sizewell A is based on a Magnox gas-cooled reactor. The much newer Sizewell B – commissioned in 1995 – incorporates a single pressurized water reactor.

LEWES, E Ssx

Birthplace, and home for some 40 years, of Gideon Mantell (1790–1852), geologist.

He had a successful practice as a surgeon and was also an avid collector of fossils: he is remembered as the discoverer of the first dinosaur to be properly described. In 1813 he moved to BRIGHTON and then sold his collection to the British Museum for £5,000, which helped him to re-establish himself in Clapham. His house at 167 High Street carries a memorial plaque. (See also WADHURST.)

LEWIS, W Is.

Callanish, the 'Stonehenge of the North', is located on Loch Roag. A circle of thirteen slabs of local granite – some weighing up to five tons – made from local stone. Dun Carloway is a fine example of a broch, the uniquely Scottish round towers dating from the Iron Age. There are other circles at Cnoc Fillibhir and Garynahine.

LEYLAND, Lancs.

The British Commercial Vehicle Museum (King St) illustrates the history of commercial road vehicles of all kinds. The exhibits are all roadworthy, and range from fire-engines to steam wagons.

LICHFIELD, Staffs.

Home of Erasmus Darwin (1731–1802) who practised medicine there for 25 years. He was keenly interested in plant taxonomy and is remembered for his *Botanic Garden*, a scientific work remarkable for being written in heroic couplets. His *Zoonomia*, or the *Laws of Organic Life*, postulated the inheritance of acquired characteristics, a view to be discredited by his illustrious grandson Charles Darwin (1809–82). There is a memorial plaque on his house by the Cathedral Close and another in the Cathedral itself.

LIMPLEY STOKE, Wilts.

The Dundas Aqueduct, an early cast-iron structure, was built by John Rennie (1761–1821) in the 1790s, his first venture into canal building.

LINCOLN, Lincs.

Robert Grossetest (*c*. 1175–1253), appointed Bishop of Lincoln in 1235, was reputedly the most learned man of his time. His tomb is in the south-east transept of the Cathedral. Like many medieval scholars he took the whole realm of knowledge as his province and wrote profusely. He wrote on optics, on comets and the generation of stars, on the reform of the calendar and much else beside of scientific significance. He was influential and caused many English Franciscans to study natural philosophy and mathematics. A much later mathematician born in Lincoln was George Boole (1815–64) who made important advances in mathematical logic at Queen's College, CORK.

Sir Joseph Banks (1743–1820) grew up in nearby Revesby. The Lawn Visitor Centre in Lincoln has a Joseph Banks Conservatory in which are grown a wide variety of the plants he brought back from his *Endeavour* voyage to the Pacific (1768–71). There is a portrait of him by Benjamin West in the Usher Art Gallery, Lincoln.

The National Cycle Museum (Brayford Wharf North) has a large collection of bicycles going back to a hobby horse of round 1820 and many cycling accessories.

The Museum of Lincolnshire Life (Burton Rd) portrays the agricultural and industrial life of Lincolnshire, especially in the 19th and early 20th centuries. Exhibits include locally manufactured farm and industrial machinery and horse-drawn vehicles.

LINDLEY, Leics.

Birthplace of Robert Burton (1577–1640). His fame rests on a single book which it took him 30 years to write. This was the *Anatomy of Melancholy* (1621), the first serious treatise on morbid psychology. (See OXFORD.)

LIPHOOK, Hants.

The Hollycombe Steam Collection has a varied collection of steam-powered machinery. It includes a sawmill, agricultural machinery and a fairground organ.

LISBURN, Down

The Giant's Ring is a prehistoric stone circle, with a tomb at its centre, nearly 200 yards in diameter. In the 18th century it was used as a race course.

LISMORE, Wat.

Lismore Castle was the birthplace of Robert Boyle (1662–91), seventh son of the 1st Earl of Cork and Orrery. His work was mostly done in England, however, because he found the Ireland of his day 'too troubled to have any Hermetic thought in it'. Famous for his discovery of Boyle's Law, relating the pressure of a gas to its volume at constant temperature. He was a founder member of the Royal Society and widely known as 'the Father of Chemistry'. (See LONDON, OXFORD.)

LIVERPOOL, Mers.

With its fortunes founded on the slave trade and the cotton industry, Liverpool became Britain's principal port for the Atlantic trade. At the turn of the century it was, with Hamburg, the principal starting point for the flood of emigrants, many of them Irish, seeking their fortunes in America. It has long had a reputation for innovation. In the latter part of the 17th century it was one of the first centres in Britain to manufacture Chinese-type porcelain. The Liverpool and Manchester Railway (see MANCHESTER), opened in 1830, was the first to handle both goods and passengers using steam locomotives. In 1842 Thomas Russell Crampton (1816–88) introduced the famous 'Crampton' locomotive. With its long boiler, and axles at the rear of the fire-box, it improved safety by having a low centre of gravity. The prototype *Liverpool*, built in 1848 for the London and North Western Railway, reached a speed of 78 mph. In Britain the class was discontinued after ten years because of damage to the track but it was enthusiastically adopted in France, so much so that for many years taking an express train was to *prendre le Crampton*. In 1863 the Liverpool-built *Banshee* was the first steel vessel to cross the Atlantic. In quite a different field, photography, The Liverpool Dry Plate and Photographic Company in 1867 paved the way to photography for the masses with the introduction of the ready-coated dry plate. In the 18th century Liverpool became an important centre for the manufacture of clocks and watches, initially as a cottage industry. In 1781 Peter Litherland patented the rack lever escapement, paving the way for cheap timepieces. In 1890 the Lancashire Watch Company was founded at PRESCOT to exploit a mass market, making watches 'so cheaply as to be in the reach of the ordinary artisan or day labourer'.

The importance of Liverpool as a port and a shipbuilding centre is reflected in exhibits in the Merseyside Maritime Museum (Albert Dock). This covers many acres and displays dry-docked ships and many marine artefacts. The Large Objects Collection (Princes Dock) displays steam vehicles, lorries, tractors, telescopes and space rockets. The basement of the Liverpool Museum in William Brown Street is devoted to transport items, from horse-drawn vehicles to locomotives. Elsewhere there are many items of scientific and technological interest, including some eighty important scientific instruments and a planetarium. There is also an exceptional collection of 18th and 19th-century clock and watch-making tools.

In the days of sailing ships vast quantities of rope were required: Nelson's *Victory* is recorded as having no less than 20 miles of rope of all calibres. Major ropewalks were located around Bold Street and Renshaw Street. The importance of the leather industry is remembered in Leather Lane and of silk in Silkhouse Court. John Cragg (1767–1854) proprietor of the Mersey Iron Foundry located in Tithebarn Street, was a pioneer of cast-iron construction, especially for churches.

Liverpool has many other personal connections with the history of science and industry. John Sadler (1720–89) invented in 1749 the transfer-printing process which made possible the rapid replication of decorated china. George Stubbs (1724–1806), best known as a painter of animals, especially horses, was born in Liverpool where his father was a currier in Ormond Street. He came to his profession by studying human and animal anatomy at YORK and in 1766 published the widely acclaimed *Anatomy of the Horse*, illustrated with his own engravings. In 1868 A.H. Saunders founded the famous manufacturing chemical firm Ayrton Saunders, now in Hanover Street: by 1920 it was employing a thousand staff. Among its products was the first 'instant' drink: tea, milk and sugar in tablet form for troops in the First World War. Sebastian de Ferranti (1864–1930) pioneer of the generation and transmission of electricity at high voltages, was born in Bold Street. Liverpool is also the home of Meccano, the famous construction toy, invented by Frank Hornby (1863–1937) – remembered also for Hornby trains – who had his works in James Street, near the station, before moving out to the suburbs.

From 1895 to 1956 Liverpool had the world's first electric overhead railway, extending 8 miles from Seaforth Sands in the north to Dingle in the south. Serving seventeen stations, it was built to relieve congestion in the dock area. Today, the most striking example of transport engineering is the Mersey Road Tunnel, opened in 1934: it is just over 2 miles in length.

Liverpool University (Brownlow Hill) received its first charter, as a university college, in 1881 and three years later was joined with Manchester and Leeds as a constituent of Victoria University. It attained independent university status in 1903. Since 1935 Liverpool has been a major centre for research in particle physics. In that year James Chadwick (1891–1974) was appointed to the Lyon Jones Chair of Physics with a specific remit to 'establish a new centre for nuclear physics research'. The appointment was appropriate, for he had experimentally demonstrated the existence of the neutron – the third subatomic particle to be identified after the electron and the positron. This discovery dramatically changed contemporary scientific views on the structure of the atom. In the same year he was awarded a Nobel Prize for Physics. The Department of Chemistry is also of international stature; the Nobel Laureate Robert Robinson (1886–1975) was Professor of Organic Chemistry 1916–21. The Liverpool School of Tropical Medicine, a logical consequence of the city's worldwide trade, also has an international reputation.

LIVINGSTON, Loth.

The Livingston Mill Farm is located in buildings which served as a working farm from the 18th century to 1960. The corn mill is driven by a 16-ft water-wheel, and is the focus of an exhibit illustrating the history of cereals and their processing.

LIZARD, Corn.

By 1900 Guglielmo Marconi (1874–1937) had succeeded in transmitting radio signals over distances up to 150 miles and was ready to attempt transatlantic communication. This was achieved on 12 December 1901 by means of a powerful transmitter at Poldhu and a receiver, operated by Marconi himself, at Cape Cod, Newfoundland. A granite obelisk, the Marconi Memorial, marks the site.

LLANBERIS, Gynd

The Welsh Slate Museum was set up in the vast Dinorwic Quarry when it closed in 1969. It shows the quarry in much the same state as it was when the last shift was worked. Exhibits include a huge water-wheel. The Museum of the North presents the history of Wales in a general way but the exhibits include ones relating to the electric supply industry.

Nearby is the terminus of the Snowdon Mountain Railway (1896), the only railway in Britain to work on a rack-and-cog system.

LLANDDANIEL-FAB, Anglesey

The impressive passage grave of Bryn-Celli-Ddu has been carefully restored. A roofed passage leads to a central chamber some 2.5 m in diameter.

LLANDOGO, Gwent

The Coed Ithel charcoal-burning furnace is a relic of ironworks established in the Wye Valley by German ironworkers in Tudor times. It is in a state of disrepair but this is an advantage in that the internal arrangement can be seen particularly well. It is comparable with the better-preserved Bonawe furnace (see TAYNUILT).

LLANDRINDOD WELLS, Powys

The Tom Norton Collection of Old Cycles and Tricycles was begun by a local bicycle dealer and repairer, Tom Norton (1870–1955). Models range from 1867 to 1965, and include penny farthings, velocipedes and tandems.

LLANDYSUL, Dyfed

The Maesllyn Woollen Mill, founded in 1881, has been in continuous operation ever since, a tribute to the high quality of its original machinery. Today, some of this is powered by restored water-wheels. There is a collection of related material.

LLANFAELOG, Anglesey

The megalithic tomb Barclodiad yr Gawres is a well-preserved passage grave, embellished with the best example of grave art outside Ireland.

LLANGOLLEN, Clwyd

In 1795 Thomas Telford (1757–1834) began work on the Pontcysyllte Aqueduct, carrying the Llangollen Canal over the River Dee. More than 300 m long, it was built on eighteen masonry piers, rising to 40 m. The canal flows through a cast-iron trough 4 m wide and 2 m deep. It took ten years to complete and is now scheduled as an Ancient Monument. It is the longest bridged aqueduct in Britain. The Canal Museum, located in a warehouse built by Thomas Telford, depicts the construction and working of canals.

LLANIDLOES, Dyfed

The Bryntail lead mine, with partially restored surface works, lies just off the B4518.

LLANRHIDIAN, Gower. See SWANSEA

LOANEND, Northld. See BERWICK-UPON-TWEED

LOCHGILPHEAD, Strath.

The area is rich in sites of archaeological interest. These include a circle and standing stones at Temple Wood and a cairn at Nether Largie, the entrance to which is embellished with ritualistic carvings.

LODE, Cambs.

In the grounds of Anglesey Abbey (NT) is an 18th-century water-mill, restored to operation in the 1980s.

LONDON

Five Centuries of Scientific Achievement

As one of the great capital cities of Europe since medieval times, London inevitably has a disproportionately rich tradition in science and technology. This was already established in Tudor times when increasingly venturesome overseas voyages of exploration and commerce engendered a demand for navigational instruments of all kinds. Instruments were required also for a wide range of other purposes, such as land survey, laying artillery, and astronomical observation. Instrument makers required not only mechanical skills but also some mathematical knowledge. Perhaps surprisingly, mathematics was then largely self-taught, for it was not much regarded in either the schools or the universities. Indeed John Wallis (1616–1703) – a founder member of the Royal Society – remarked much later that 'the study of mathematicks was cultivated more . . . in London than in the universities.' Again, John Newton (1622–78), who taught at Ross-on-Wye, declared 'I never yet heard of a Grammar School in which mathematics is taught . . . and there are not many Tutors in either of the Universities that do.' One London teacher we can identify is Edward Cocker (1631–75), who taught writing and arithmetic by St Paul's Churchyard. His *Arithmetick, being a Plain and Easy Method* appeared in 1678. It ran through over a hundred editions and influenced the teaching of mathematics in England for more than a century. Very little is known about him but his name is remembered wherever the English language is spoken: 'according to Cocker' is synonymous with something that can be relied upon.

One of the earliest of the instrument-making fraternity was Nicholas Kratzer (1486–1550) of Munich, who in 1519 was appointed astronomer and horologist to Henry VIII, who later sent him to Oxford to teach astronomy and 'the doctrine of the spheres'. Many of his instruments, especially sundials, survive. More closely linked with London was Thomas Gemini (*fl.* 1524–62), whose workshop was 'within the precincts of the late Blackfriars'. He first made his name by engraving the magnificent plates for the English edition (1545) of Vesalius' *Anatomy*: from that it would be a natural transition to engraving the scales of scientific instruments, a task demanding great accuracy. Probably among his pupils was Humfrey Cole (1530?–91), another prolific and skilled instrument maker who described himself as a north countryman 'pertaining to the Mint in the Tower' with a workshop 'near the north door of St Paul's'.

The versatile Cambridge mathematician John Dee (1527–1608) had a keen interest in the practical applications of astronomy and geometry and the design of instruments. After some years on the Continent he was introduced to Court circles in London, where he survived, though with varying fortune, for some 30 years: he was adviser to several important British voyages of discovery. His dabbling in alchemy, astrology and the occult somewhat clouded his reputation but nevertheless the great Danish astronomer Tycho Brahe described him and Thomas Digges as 'most noble, excellent, and learned mathematicians'. Thomas Digges (*c.* 1543–95), the son of Leonard Digges (1510–58), an authority on navigation, was also an instrument maker and a close friend of Dee. Very recently the two Digges have sprung to new fame, for convincing evidence has been uncovered that the telescope – supposedly invented in Holland in 1608 – was in fact invented by them in London 30 years earlier.

Generally associated with Dee and Thomas Digges as one of the three leading mathematicians in England was Thomas Har(r)iot (1560–1621). After graduating at Oxford he went into the service of Walter Ralegh, at Durham House in the Strand, who promptly dispatched him to Virginia as mathematical adviser to his colonists there. After Ralegh was sent to the Tower in 1603, Hariot entered the service of the Earl of Leicester at Sion House, Isleworth. There, as early as June 1609, he made some of the very first telescopic observations of the Moon and saw 'Venus horned like the Moon, and spots on the Sun'.

But perhaps the most famous of all instrument makers of this period was Elias Allen (*fl.* 1606–54) whose workshop was at the Sign of the Horse Shoe in the Strand. He made fine, richly ornamented instruments in brass and silver, of which many survive. The business addresses of these craftsmen have a charm of their own. Christopher Jackson (*fl.* 1590–1616), who specialized in surveyors' measuring-chains, operated at the Sign of the Cock, Crooked Lane, Eastcheap. Ralph Greatorex (1625–1712) was to be found at the Sign of Adam and Eve in the Strand.

However, the existence of so many craftsmen is not to be taken as evidence of organized science in London at that time. The instrument makers no doubt knew each other, and many certainly trained apprentices to carry on the trade, but in the main they seem to have operated individually. Greatorex was once referred to as the 'Doyen of the Mathematical Instrument Makers' Club', but of this nothing is known.

For anything approaching science as we conceive it today, we must look at Gresham College (Gresham Street, EC2). This was endowed in his will by Sir Thomas Gresham (1519–79) – the Elizabethan financier who built the first Royal Exchange – for the delivery of public lectures (in both Latin and English) on astronomy, geometry, physics, divinity, music, law and rhetoric for the benefit of both the citizens of London and foreign visitors. News of his intention seems to have reached Oxford and Cambridge, each of which considered their own cities as more suitable sites than London. If so, Gresham was unmoved.

The lectures were well attended, so much so that Thomas Sprat (1635–1713), in his *History of the Royal Society* (1667), remarked – somewhat patronizingly – that 'if it were beyond the Sea, it might well pass for a *University*'. The foundation of the College was timely, for it provided a focus for the talented, but hitherto largely unorganized, scientific community in London. A major scientific event at just that time was the publication in 1600 of *De magnete* by William Gilbert (1544–1603) of COLCHESTER, the first classic of modern physics. He had been appointed physician to Elizabeth I and living in London since 1573. As early as 1616, William Harvey (1578–1657) had been expounding his, as yet unpublished, revolutionary views on the circulation of the blood. Gilbert's friend Henry Briggs (1561–1631) was appointed first Professor of Geometry at the College and was busy disseminating understanding of John Napier's logarithms. His colleague, as Professor of Astronomy, was Edmund Gunter (1581–1626), also a keen advocate of logarithms and inventor of several scientific instruments. His successor Henry Gellibrand (1597–1637) completed Briggs' plan to prepare very accurate trigonometrical tables based on logarithms. All three were much concerned with the practical application of their work in navigation and astronomy and were in close touch with officers and administrators of the Royal Navy. A regular visitor to the College was William Oughtred (1575–1660), rector of ALBURY in Surrey, perhaps the most outstanding mathematician of his generation in England; among his achievements was the invention of the slide-rule. Among others associated with the London scientific community in the first half of the 17th century were Christopher Wren (1632–1723), mathematician and architect; John Wallis (1616–1703), a mathematician ranked with Oughtred; and the young Robert Boyle (1623–91), the future 'Father of Chemistry'. Francis Bacon (1561–1626) was pursuing his concept of 'a grand instauration of the sciences' within which all knowledge would be comprehensively reviewed for the benefit of mankind. Their domestic discussions were, of course, stimulated by news of startling events on the Continent. Galileo's astronomical observations with the newly available telescope, published in his slim *Siderius Nuncius* (1610) suddenly revealed hitherto unsuspected details of the Moon, planets and outer space. Twenty years later his *Dialogo* (1632), refuting the Ptolemaic system of the Universe in favour of the Copernican, involved him in near-fatal dispute with the Inquisition. Galileo's pupil Torricelli (1608–47) demonstrated by a very simple, but very dramatic, experiment the error of the Aristotelian belief that 'nature abhors a vacuum'. In France, René Descartes (1596–1650) in his *Géometrie* (1637) had forged a powerful new weapon for applying algebraic methods to geometry, with important implications for the solution of scientific problems.

The lectures at Gresham College informally brought together most of London's scientific community. From about 1645, however, some of them conceived the idea of regular weekly meetings in the College, or near by, to discuss the scientific topics of the day. This development was interrupted by the Civil War. Around 1648/9, a number of the group moved to OXFORD, where they continued to meet informally with like-minded members of the University. Meanwhile, meetings continued at Gresham College but rather little is known about them except that they became more frequent in the last years of the Protectorate.

The Restoration brought about a development of exceptional importance for the evolution of British science. At a meeting of the College on 28 November 1660 it was resolved formally to establish a scientific society. Very quickly, some forty people were judged 'willing and fit to joyne them'. Within a week the King had signified his approval and support – not surprising as the membership was strongly Royalist – and in 1662 granted a Royal Charter. Moreover, 'hee was pleased to offer of him selfe to be enter'd one of the Society'.

Thus was born The Royal Society of London for 'further promoting by the authority of experiments the sciences of natural things and of useful arts'. It was by no means the world's first scientific society but despite varying fortunes over the centuries it grew to be one of the most influential and prestigious. Until the advent of Nobel prizes at the beginning of this century Fellowship was the highest honour to which a British scientist could aspire and the Presidency was the highest accolade of all. Until 1710 the Society continued to meet at Gresham College: this was demolished in 1768. It has had several moves since – including Somerset House, Strand, and Burlington House, Piccadilly – and since 1967 has been housed in palatial premises in Carlton House Terrace, the pre-war German embassy.

The foundation of the Royal Society was important not only in itself but also because it corresponded with a new era in science – the final abandoning of Aristotelian dogma and the pursuit of knowledge by experiment and rigid argument. In 1684 Newton began to compose his great *Principia*, which he presented to the Society in 1686: perhaps the most profoundly influential scientific work of all time, it was published in the following year.

It was not only in science that London was an important focal point. Although much industry was located there, it was never a great industrial centre in the sense that BIRMINGHAM, LIVERPOOL, MANCHESTER, and the other great conurbations of the Industrial Revolution were destined to be. Nevertheless, the city played a central role in the development of new sources of power.

Apart from animal and manpower – as in treadmills and oared boats, for example – the sole sources of power until the beginning of the 18th century were wind and water. In 1698 Thomas Savery (1650–1715), an engineer familiar with the problems of mine drainage in Cornwall, patented an engine for raising water 'by the impellent Force of Fire'. In fact it was not so much an engine, in the sense that it had no moving parts, as a water pump based on the cyclical condensation of steam. He established a workshop in London, but his engines had little success. Nevertheless he was elected a Fellow of the Royal Society in 1705 and appointed superintendent of the waterworks at Hampton Court. Meanwhile, however, another West Country man with an interest in mine drainage, Thomas Newcomen (1663–1729), had devised a clumsy but reliable steam engine incorporating a great pivoted beam. The first of these was erected in 1712 at Dudley, Worcs., and subsequently large numbers were constructed, especially in the developing coal mines and in the metal mines of Cornwall. Newcomen's engine was totally different from Savery's pump and invented quite independently of it. Nevertheless, it was undeniably dependent on 'the impellent Force of Fire' and thus infringed Savery's patent, which eventually ran to 1733, and Newcomen had to come to an accommodation with him. This obliged him to spend time in London. Newcomen died in London, and it is possible that it was his home for the last years of his life.

The story does not end there. In 1764 James Watt (1736–1819), in GLASGOW, was asked to repair a model Newcomen engine and this led him to a brilliant improvement, the introduction of a separate condenser. From this emerged a new generation of steam engines, manufactured from 1774 by the famous partnership of Boulton and Watt in BIRMINGHAM. These were to be the main power units of the latter part of the Industrial Revolution.

London was also closely identified with another important, but quite different source of power. At the beginning of the 19th century there appeared in London a flamboyant German company promoter, Frederick Winsor (1763–1830), with extravagant plans to promote the use of coal-gas for lighting, earning vast profits for investors. In 1807 he achieved great publicity by illuminating with coal-gas a wall of Carlton House to celebrate the King's birthday. Ultimately, the London Gas Light and Coke Company was founded in 1812. Gaslight was quickly a huge success for lighting streets, factories and homes and by 1830 some two hundred gas companies had been founded up and down the country. It received a great boost in 1821 when the Prince Regent had it installed in his Pavilion at BRIGHTON. Winsor profited little, however, for his personality proved more of a liability than an asset to the new company and he was quickly dropped. Nevertheless, he was a man of vision: he was the first to realize that the secret of success with gas was distribution from a central source rather than local generation. Later, from the middle of the 19th century, coal-gas was used for the first effective internal combustion engines, from which derived the petrol, and later diesel, engines which revolutionized transport in the 20th century.

London was the birthplace of yet another important source of power, electricity. Electrical effects – such as the ability of rubbed amber to attract fragments of dried leaves – had been known from antiquity. Its experimental study dates from the mid-17th century, but not until Volta invented the battery in 1799 was it possible to obtain a continuous flow of electricity. In London the voltaic cell was used to great effect at the Royal Institution by Humphry Davy (1778–1829) to isolate several new elements, and by Michael Faraday (1791–1867) whose recognition in 1831 of the relationship between electricity and magnetism led to the invention of the dynamo, destined to be the basis of a vast international electricity industry.

In the early 18th century, however, electrical phenomena were no more than a laboratory curiosity, of no practical consequence. What was occupying the public interest was the possibility of making huge fortunes by speculation in commercial voyages overseas. The South Sea Company was founded in 1711 to trade, mainly in slaves, with Spanish America. When the Bubble burst many investors were ruined and three ministers of the crown were charged with corruption. This may seem far removed from contemporary science, but there was in fact a close relationship. Whatever the motivation, navigation on the high seas depended on ability to determine a ship's precise position. Latitude presented no difficulties as – given clear skies – it could be determined by stellar observation. Longitude, however, was a very different matter: in essence, it depends on knowing local time relative to that at some agreed longitude (in practice, that of Greenwich). While the best pendulum clocks by then kept good time on land, they were useless on a ship pitching and rolling at sea. So important had the problem become – for both naval and merchant shipping – that in 1714 the British government set up a Board of Longitude, empowered to award a prize of £20,000 – a huge sum in those days – to anybody devising a practical solution. The answer was eventually found by John Harrison (1693–1776) whose famous No. 4 chronometer (now preserved at GREENWICH) in 1761 lost only five seconds after a nine-week voyage to Jamaica. Nevertheless, the Board – in the manner of official bodies – gave only grudging approval and not until 1773 was the award paid in full. It can be argued that the marine chronometer did for navigation in the 18th century what the magnetic compass did in the 12th century and radar did in the 20th. 'Longitude' Harrison, as he was called, died at his home in Red Lion Square WC1, and is buried in Hampstead Church.

The solution of the longitude problem was a striking demonstration of the great practical value of science. This was recognized in a more general way by the foundation in 1754 of the Society for the Encouragement of Arts, Manufacture and Commerce, now the Royal Society of Arts.

The 19th century saw a great expansion in what may be called the scientific infrastructure of London. In 1799 the cosmopolitan Benjamin Thompson, Count Rumford (1753–1814) – who married Lavoisier's widow – had founded the Royal Institution in Albemarle Street to promote science by lectures and research. One consequence of this was the invention of the miner's safety lamp by Davy in 1812. University College – 'the godless college in Gower Street' – was founded in 1826 to provide university education for non-Anglicans, excluded from Oxford and Cambridge. There, in 1894, Sir William Ramsay (1852–1916) made history by discovering a whole new family of elements, the rare gases of the atmosphere. For this he received one of the first Nobel prizes (1904). Its godly rival, King's College, in the Strand, was founded three years later. There, in the 1840s, Charles Wheatstone (1802–75) and William Cooke (1806–79) carried out their pioneer experiments on the development of the electric telegraph. Joseph Lister, who introduced antisepsis into surgery, was professor there 1877–92. In mid-century three scientific and technical colleges were founded in South Kensington: in 1907 they were amalgamated to form what is now the internationally famous Imperial College of Science and Technology.

Two other highlights of scientific endeavour, in quite different fields, deserve mention. William Wollaston (1766–1828) succeeded in 1805 in mastering the recalcitrant element platinum – valuable because of its exceptionally high resistance to corrosion – and producing it in malleable form. His laboratory, where his research was done in great secrecy, was at 14 Buckingham Street, WC2. Charles Babbage (1792–1877), pioneer of machine computing, did most of his important work while living in London. His famous difference engine is preserved in the Science Museum.

The central event of 19th-century London – both in time and public interest – was undoubtedly the Great Exhibition of 1851. Staged in Joseph Paxton's huge Crystal Palace in Hyde Park, under the patronage of Prince Albert, it was an international display of industrial achievement, from cutlery to steam engines, carpets to carriages and locomotives. It was visited by over six million people, a number made possible mainly by the availability of cheap railway fares. The exhibits were judged by an international jury and Britain swept the board, seemingly justifying her claim to be the Workshop of the World. But a severe shock was in store: at the Paris Exhibition of 1867 she gained no more than a dozen awards. Inquiry indicated that the biggest single cause was Britain's failure, *vis-à-vis* her European competitors, to evolve an education system appropriate for the needs of the day. When the Exhibition closed the Crystal Palace was removed to Sydenham, only the gates remaining in Hyde Park.

The dawn of the 20th century was also the dawn of a new era in communication. In 1901 Marconi caused a sensation with his first transatlantic radio signal. His apparatus was primitive, but a major step forward was made with the invention in 1904 of the thermionic valve by J.A. Fleming (1849–1945), Professor of Electrical Engineering in University College, London. This was an essential component of virtually all radio transmitters and receivers, and of a wide range of other electronic devices such as television sets, until challenged by the transistor after the Second World War. It opened the way to public broadcasting: the British Broadcasting Corporation began its service in 1927. A year earlier John Logie Baird (1888–1946) had demonstrated his photomechanical television system in Soho and in 1929 the BBC began a regular programme, the world's first, on five mornings a week. In the event, however, Baird's system was not on the main line of progress and in 1936 it was displaced by the present all-electric system.

London has long been famous for its great teaching hospitals and in one of these – St Mary's, Paddington – one of the greatest medical discoveries of all time was made in 1928. In that year Alexander Fleming (1881–1955) made his serendipitous, but none the less historic, discovery of penicillin. He himself did not recognize its unique therapeutic powers but ten years later it was developed at OXFORD and, through a powerful Anglo-American industrial programme, became available to all Allied military casualties during and after the D-Day landings in Normandy. Not until after the war was it available for civilian use. Another important discovery in the medical field was that of interferon in 1957 by Alick Isaacs (1921–67) at the National Institute of Medical Research, Mill Hill.

As befits a great capital city, London possesses many fine buildings and monuments, and among these three civil engineering achievements are outstanding. Tower Bridge, completed in 1894, is London's best-known bridge and a famous landmark. In 1964 the city acquired an equally prominent new landmark. This is the London Telecom Tower, near Tottenham Court Road. Much more recent is the huge Thames Barrier, completed in 1982. Its purpose is to prevent the disastrous floods which have plagued London for centuries. But another great civil engineering achievement is not immediately apparent. The world's first underground railway, from Paddington to

Farringdon Street, was opened in 1863 and was steam-operated: the introduction of electric traction in 1890 vastly improved conditions for passengers. Today, London Transport trains operate on over 250 route-miles, of which 100 miles are underground.

In this century science has expanded enormously, both nationally and internationally. At the same time it has become more diffuse, research being pursued by groups rather than individuals and at a number of centres. In London, as elsewhere, the sharply defined individual achievements of the past are tending to be subsumed within a general advance on many fronts.

Some Resident Scientists and Engineers

All communities take pride in honouring their distinguished members, be it a single one in a small town or many in a large city. London is no exception and the observant visitor will see countless memorials – statues, busts, wall plaques and the like – to famous people who in one way or another have been associated with the city. But there is one place exceptionally rich in such memorials: this is Westminster Abbey, consecrated in 1065, where over a thousand names are memorialized. Among them are more than fifty scientists and engineers, many already encountered in this essay. They range from Geoffrey Chaucer (1343–1400), an accomplished astronomer as well as author of the *Canterbury Tales*, to Lord Florey (1898–1968) whose 'vision, leadership and research made penicillin available to mankind'. The last scientist to be so honoured was Paul Dirac, in 1996. (See BRISTOL.) In between are such giants as the mathematicians Isaac Newton and Isaac Barrow; Stephen Hales, physiologist and botanist; Lord Lister, pioneer of antiseptic surgery; the engineers George and Robert Stephenson and James Watt; Sir Charles Parsons, inventor of the steam turbine; Charles Darwin and Alfred Russel Wallace, pioneers of the theory of organic evolution; the astronomers William and John Herschel; James Clerk Maxwell, mathematical physicist; Humphry Davy and Michael Faraday of the Royal Institution; William Ramsay, discoverer of the rare gases of the atmosphere, and J.J. Thomson and Lord Rutherford, who laid the foundations of atomic physics. The memorials take various forms, from a bust to a window, from a medallion to a slab in the floor.

The Abbey is thus a splendid place for those who wish to pay homage to the great without too much legwork. For the more peripatetic there are a host of blue commemorative plaques attached to buildings – initially by the Royal Society of Arts in 1867 but now the responsibility of English Heritage – and many public statues and other memorials. These are far too numerous to be listed here: only some of the most conspicuous ones can be noted. They do not, of course, mean that the people concerned were true Londoners, but merely that they were residents for varying periods of time.

Rudolf Ackermann (1764–1834) owned a fashionable book and print shop at 101 The Strand, WC2. It was the first shop in the world to be lit by gas. He also invented the Ackermann steering system for road vehicles.

Sir Richard Arkwright (1732–92), the industrialist primarily responsible for the mechanization of the textile industry, lived for a time at 8 Adam Street, WC2.

John Logie Baird (1888–1946) first demonstrated television in 1926, in his house at 22 Frith Street, W1. There is a pub named after him in Muswell Hill.

Sir Joseph Banks (1743–1820), botanist who accompanied James Cook on the

Endeavour voyage of scientific exploration 1768–71 and President of the Royal Society 1778–1819, lived at 32 Soho Square, W1 (now rebuilt).

Sir Joseph Bazalgette (1819–91), civil engineer who designed and built a new drainage system for London 1855–89, lived at 17 Hamilton Terrace, NW8. There is a bronze bust on the Victoria Embankment, EC4.

George Bentham (1800–84), plant taxonomist, lived at 25 Wilton Place, SW1.

Richard Bright (1789–1858), physician who identified the form of nephritis that bears his name, lived at 11 Savile Row, W1.

Sir Marc Isambard Brunel (1769–1849) and his son **Isambard Kingdom Brunel** (1806–59), engineers, lived at 98 Cheyne Walk, SW10. Both are buried in Kensal Green Cemetery. There are statues of I.K. Brunel on the Embankment, and at Paddington Station.

Henry Cavendish (1731–1810), eccentric chemist who first discovered hydrogen ('inflammable air'), lived for many years at 11 Bedford Square, WC1.

Sir George Cayley (1773–1857), pioneer of aviation theory, lived at 20 Hertford Street, W1.

Captain James Cook (1728–79), commander of the first scientific voyage of discovery to the Pacific (*Endeavour* 1768–71) lived at 88 Mile End Road. His statue is in The Mall, SW1, near Admiralty Arch.

Sir William Crookes (1832–1919), famous for his discovery of thallium and research on cathode rays, lived at 7 Kensington Park Gardens, W11.

Sir Frank Dyson (1868–1939), Astronomer Royal 1910–33, lived at 6 Vanbrugh Hill, SE3. He organized the observation of the solar eclipse of 1919 at which Einstein's predicted gravitational deflection of light passing the Sun was experimentally detected.

Thomas Earnshaw (1749–1829), famous clock and watch-maker, had his workshop at 119 High Holborn, WC1.

Sir Arthur Eddington (1882–1944), astrophysicist, lived at 4 Bennett Park, SE3.

Michael Faraday (1791–1867), famous for his experiments on electromagnetism at the Royal Institution, was apprenticed to a bookbinder at 48 Blandford Street, W1. A statue stands outside the Institution of Electrical Engineers, Savoy Place, WC2.

Sir Alexander Fleming (1881–1955), discoverer of penicillin, lived at 20a Danvers Street, SW3. There is a memorial window in St James' Church, Paddington: also a local pub named after him.

Sir Ambrose Fleming (1849–1945), inventor of the thermionic valve, lived at 9 Clifton Gardens, W9.

Benjamin Franklin (1706–90), American statesman/scientist, lived at 36 Craven Street, WC2.

William Frieze-Green (1855–1921), pioneer of cinematography, lived at 136a Maida Vale, W9.

Sir Francis Galton (1822–1911), statistician and founder of eugenics, lived nearly his whole life at 42 Rutland Gate, SW7.

James Glaisher (1809–1903), astronomer and pioneer meteorologist, lived at 20 Dartmouth Hill, SE10.

Henry Gray (1827–61), author of the best-known student's book on anatomy, lived at 8 Wilton Street, W1. This was the standard work, in many editions, for more than a century.

John 'Longitude' Harrison (1693–1776), inventor of the marine chronometer for the accurate determination of longitude at sea, lived in a house on the site of Summit House, Red Lion Square, WC1.

John Hunter (1728–93), outstanding surgeon and anatomist, lived at 31 Golden Square, W1. There is a bust in Leicester Square.

William Hunter (1718–87), perhaps the greatest of all teachers of anatomy, founded a famous school of anatomy in Great Windmill Street, W1, in 1770.

T.H. Huxley (1825–95), a protagonist of Darwin's theory of evolution (Darwin's Watchdog), lived at 38 Marlborough Place, NW8.

John Innes (1829–1904), founder of the world-famous Horticultural Institute, ROTHAMSTED, lived at Manor House, Watery Lane, SW20.

Edward Jenner (1749–1823), pioneer of smallpox vaccination, is commemorated by a statue in Kensington Gardens. (See BERKELEY.)

Lord Lister (1827–1912), pioneer of antiseptic surgery, came to London from EDINBURGH in 1877, and lived at 12 Park Crescent, W1.

Sir Patrick Manson (1844–1922), famous for his research on insect-borne diseases, especially malaria, lived at 50 Welbeck Street, W1.

Guglielmo Marconi (1874–1937), Italian pioneer of radio communication, lived at 71 Hertford Street, W1.

James Clerk Maxwell (1831–79), professor at King's College 1860–5 and creator of the electromagnetic theory of light, lived at 16 Palace Gardens Terrace, W8.

Sir Isaac Newton (1642–1727), among the greatest mathematicians of all time, came to London from CAMBRIDGE in 1699 as Master of the Mint. For a time he lived in a house on the site of what is now Westminster Public Library, St Martin's Lane, WC2; the one designated by a plaque is 87 Jermyn Street, SW1. He is also commemorated by a stone head in Leicester Square, WC2.

Samuel Pepys (1633–1703), diarist and one of the earliest Fellows of the Royal Society (1604) lived in a house now replaced by 12 Buckingham Street, WC2.

Lt-Gen. Augustus Pitt-Rivers (1827–1900), anthropologist, archaeologist and founder of the Pitt-Rivers Museum, OXFORD, lived at 4 Grosvenor Gardens, SW1.

William Plimsoll (1824–98), 'the sailor's friend' and originator of the Plimsoll line for ships (1844), is represented by a bronze bust in Victoria Embankment Gardens, WC1. (See BRISTOL.)

Joseph Priestley (1733–1804), the chemist, is depicted as a seated figure over the entrance to the Royal Society of Chemistry premises in Russell Square.

Sir Ronald Ross (1857–1932), famous for his research on malaria for which he was awarded one of the earliest Nobel prizes (1902) lived at 18 Cavendish Square, W1.

Sir Hans Sloane (1660–1753) was the only person to have been simultaneously President of the Royal College of Physicians (1719–35) and the Royal Society (1727–41). His great collection was the basis of the British Museum, where his statue stands. He is portrayed on the sign of the nearby Museum Tavern in Great Russell Street. He lived at 41 Bloomsbury Place, WC1, and is buried at Chelsea Old Church.

Robert Stephenson (1803–59), engineer, lived and died at 35 Gloucester Square, W2. His statue stands in the forecourt of Euston Station.

John Tradescant (d.1638), naturalist and gardener to Charles I, established a physic garden and museum about 1613, on the east side of South Lambeth Road. The curios he collected on his extensive foreign travels formed the nucleus of the museum established by his friend Elias Ashmole at OXFORD in 1682.

Richard Trevithick (1771–1833), pioneer of steam traction and high-pressure steam, is remembered in a bronze tablet at University College, London.

Thomas Wakeley (1795–1862), founder of the world-famous medical journal *The Lancet*, lived at 35 Bedford Square, WC1.

Alfred Russel Wallace (1823–1913), co-founder with Charles Darwin of the theory of organic evolution, lived at 6 Tressillian Crescent, SE4.

Chaim Weizmann (1874–1952), organic chemist and first President of Israel, lived at 67 Addison Road, W14. (See MANCHESTER.)

Sir Henry Wellcome (1853–1936), drug manufacturer and founder of the Wellcome Trust, lived at 6 Gloucester Gate, NW1.

Thomas Young (1773–1810), founder of the wave theory of light and interpreter of the hieroglyphics on the Rosetta Stone, lived at 48 Welbeck Street, W1.

Historic Buildings and Sites

If one structure had to be singled out as epitomizing the London scene it must surely be Tower Bridge, which is to the city what the Eiffel Tower is to Paris. Impressive in itself, it is made more so by its proximity to the Tower, with HMS *Belfast* (the responsibility of the Imperial War Museum) moored close by. Its unusual design was determined by its position so far downstream, for provision had to be made for the free passage of large ships. It is a bascule bridge, the deck being divided at the centre and hinged at each end so that the two parts can be made vertical: this gives a clear width of 62 m. A footbridge 42 m above the water can be used by pedestrians when the bridge is open (now rarely). The engineer was Sir John Wolfe-Barry (1836–1918) and it was completed in 1894. The original hydraulic lifting gear was preserved after electrification in 1976. There is a small associated museum.

The 1660s was a memorable decade for London, for it saw the Great Plague (1664/5), the Great Fire (1666) and the Great Flood (1663). Pepys recorded in his *Diary* for 7 December 'there was last night the greatest tide that ever was remembered . . . in this river, all Whitehall having been drowned'. Other calamities followed: as recently as 1928 fourteen people drowned when basements were flooded at Westminster. This led to the building of the Thames Barrier between Greenwich and Woolwich: it was completed in 1982. Its span is 500 m and associated embankments run some 30 km downstream. It is designed to contain flood water rising 7 m above the normal level. It is a dramatic engineering project by world standards. There is a visitors' centre at Unity Way, SE18, at which various displays illustrate the construction of the Barrier and how it is operated.

Another outstanding landmark of recent origin is the British Telecom Tower, centre of BT's world-wide telecommunications network. It soars nearly 200 m above Cleveland Street, W1.

In London, memorial plaques are not limited to the homes of the famous: some sites of particular interest are also so designated. One at 16 Lawrence Street, SW3, recalls the old fame of the city as a centre for pottery manufacture: 'On this site Chelsea pottery was manufactured 1745–84'. At Alexandra Palace, Muswell Hill, N22, the BBC erected their first television transmitter: the first transmission, of 'Here's Looking at You', was made on 26 August 1936. Much of the building was destroyed by fire in 1980. A plaque in Westferry Road, E14, marks the shipyard where the *Great Eastern* was launched in 1858. Not far away, in Canning Town, E16, was the Thames Ironworks and Shipbuilding Company, founded in 1856. By 1871 it was described as 'the greatest shipyard of all'. It was famous for the construction of HMS *Warrior* (1860), an armoured frigate then the largest warship afloat: she is now berthed at PORTSMOUTH. The works also made castings for Hammersmith Bridge, the roof of the Alexandra Palace and the Menai Bridge. Another designated site is the railway viaduct at Walthamstow Marshes, E17. Here A.V. Roe (1877–1958) assembled his No. 1 triplane and in July 1909 made the first powered flight in Britain. The workshop of Sir Hiram Maxim (1840–96), pioneer of the machine-gun, was situated at what is now 57d Hatton Garden, EC1.

In Britain, church bells were commonly cast near the church for which they were destined: hence the number of Bell Fields dotted around the countryside. But in Whitechapel Road, E1, is a national bell foundry that dates back to 1420: it was then located in Houndsditch, moving to its present site in 1738. Many famous bells have been cast here, including Big Ben, the original Liberty Bell in the USA, and the Bicentennial Bell presented by Britain to the American people in 1976. Almost as common as Bell Fields in south-east England are high points designated as Telegraph Hill. These are reminders of mechanical telegraph systems, using semaphore arms, introduced by the Admiralty during the Napoleonic Wars to link London with the main naval ports. The Telegraph Inn, Putney Heath, is one of many reminders of this once important system of communication.

The middle of the 19th century saw a tremendous increase in rail transport and three of London's great main-line stations – now listed as Grade I buildings – were built then by rival companies: King's Cross (1852), Paddington (1854) and St Pancras (1867). All are magnificent examples of Victorian construction. Queen Victoria arrived at the original Paddington Station after her first railway journey, from Slough, in 1842. This was replaced by the present one, designed by the Great Western Railway's engineer Isambard Kingdom Brunel (1806–59). In its extensive use of cast and wrought iron and glass, he was much influenced by the Crystal Palace, designed by Sir Joseph Paxton (1801–65) for the Great Exhibition of 1851. The central aisle has a span of 40 m. Initially it was served only by Brunel's beloved broad-gauge (7 ft ¼ in) track, but this was finally abandoned in 1892, with mixed gauge in use from 1861. In 1954, to mark the station's centenary, a plaque in memory of Brunel was placed on the wall by Platform 1.

The first tunnel under the Thames – 1 km from Rotherhithe to Wapping – was completed in 1843 by Marc Isambard Brunel (1769–1849): originally a thoroughfare, it was converted to railway use in 1871. The remains of his engine house still stands in Tunnel Road, Rotherhithe, SE16. A pumping station, designed by the well-known water engineer Thomas Hawksley (1807–93) in 1858, was located in what is now the pavilion at the north end of Kensington Gardens, W8.

In London, potable piped water was still not generally available even in the middle of the 19th century, by which time many wells were becoming seriously contaminated, leading to epidemics of water-borne diseases, such as cholera. Such diseases were commonly believed to be caused by some sort of atmospheric miasma but in 1849 John Snow (1813–58), obstetrician to Queen Victoria, correctly diagnosed that they were caused by contaminated public water supplies. The pump in Broad Street was particularly suspect, and Snow's point was proved when the authorities removed the pump-handle and the local incidence of cholera dropped dramatically. Broad Street is now Broadwich Street, Soho, W1, and Snow's name is commemorated by a local pub.

At the end of this catalogue of places in London with scientific connotations perhaps the most historically significant of all, but quite insignificant in appearance, is the junction of Kingsway and High Holborn, WC2. It was there – as recorded in Robert Jungk's *Brighter than a Thousand Suns* – that the refugee atomic physicist Leo Szilard (1898–1964), while waiting for the traffic lights to change so that he could cross, first conceived the idea that it might be possible to use neutrons to break up atomic nuclei and initiate chain reactions releasing vast amounts of energy. Later he moved on to the USA and was associated with Enrico Fermi (1901–54) with the production of the world's first chain reaction – in a squash court in the University of Chicago – on 2 December 1942. Thus was born the atomic bomb.

Museums of Science, Engineering and Medicine

London is naturally the home of the great national collections, notably, in the present context, the Science Museum and the Natural History Museum. The third great museum – the British Museum in Bloomsbury, from which the Natural History Museum was hived off in 1881, mainly to house the collections of Sir Joseph Banks and Sir Hans Sloane – is very broadly based but holds a large amount of material of scientific interest, especially time-keepers and scientific instruments. Additionally, there are many smaller museums which are devoted wholly or substantially to displaying material of scientific, engineering, or medical interest. The selection listed below includes most of those that are open to the public, at least at times. Many charge an admission fee. It excludes those which, for various reasons, can be visited only by special arrangement, usually by those with a demonstrable professional interest.

BT Museum, 145 Victoria Street, EC4. Exhibits illustrate the history of telecommunications, primarily electrical but including some early mechanical devices such as the semaphore.

The Worshipful Company of Clockmakers, Guildhall Library, Aldermansbury, EC2. It exhibits a fine representative collection of timepieces by leading British makers.

The Faraday Museum, 21 Albemarle Street, W1. This is located in the Royal Institution, where Faraday carried out his classic research on electromagnetism. It displays some of his apparatus, papers, personal effects and other memorabilia.

Museum of Garden History, St Mary-at-Lambeth, Lambeth Road, SE1. The Tradescant Trust has turned this church – in which are buried two famous gardeners, John Tradescant (1570–c. 1638) and his son, also John (1608–62) – into a museum devoted to all aspects of garden history, including tools. Also buried here is Elias Ashmole, who acquired the Tradescant collection of curios as the basis of the Ashmolean Museum in OXFORD.

The Heritage Motor Museum, Brentford. This has an extensive collection of British motor-cars, including many famous marques.

Horniman Museum, 100 London Road, SE23. This small museum of natural history displays a wide range of stuffed animals.

Imperial War Museum, Lambeth Road, SE1. This was conceived in the last year of the First World War and founded in 1920 to keep a permanent record of British military memorabilia relating to the war. It has since been greatly extended, especially in the aftermath of the Second World War, and is now a very large and impressive national collection. Apart from military hardware – including many very large items – the material includes many paintings, photographs, film and other related items. The museum also administers the aeronautical museum at DUXFORD airfield and HMS *Belfast* moored in the Pool of London by Tower Bridge.

Kew Bridge Steam Museum, Green Dragon Lane, TW8. This depicts the history of public water supply but the most dramatic exhibits are the old pumping engines, some very large.

London Canal Museum, New Wharf Road, N1. This is housed in an old warehouse beneath which was stored, in the 19th century, ice imported from Scandinavia. Exhibits include canal boats and memorabilia related to inland waterways.

London Planetarium, Marylebone Road, NW1. This is located next door to the famous Madame Tussaud's Waxwork Museum (which includes representations of Christopher Wren and Benjamin Franklin). The domed planetarium has a fine Zeiss projector, capable of providing a wide range of programmes.

London Transport Museum, Covent Garden, WC2. This portrays transport in London from horse buses to electric trams, trolley-buses, and underground rolling-stock.

Museum of the Moving Image, South Bank, SE1. Devoted, as its name implies, primarily to the history of cinema and television. The exhibits include also many precursors of cinematography designed to give the illusion of movement, such as the zoetrope and the bioscope.

National Army Museum, Royal Hospital Road, SW3. This museum of military technology is concerned particularly with the evolution of small arms from the 16th century. There are also associated exhibits of instruments such as heliographs and range-finders.

National Maritime Museum, Greenwich, SE10. This is one of the world's greatest museums devoted to the history of ships and seamanship, including a fine collection of model ships and navigational instruments. Also at Greenwich the Cutty Sark Trust preserves historic British ships from the *Cutty Sark* itself to the *Gipsy Moth IV* in which Francis Chichester circumnavigated the globe single-handed in 1966/7.

The Natural History Museum, Cromwell Road, SW7. This internationally famous museum, library and research centre occupies the huge ornate building designed by Alfred Waterhouse (1830-1905) in the 1870s to house the natural history collections in the British Museum. It contains tens of millions of items – not all, of course, on display – ranging from arthropods to dinosaurs and a blue whale. It also administers the Zoological Museum, TRING.

North Woolwich Old Station Museum, E16. The station, built in 1854 to serve the Great Eastern Railway ferry across the Thames, was closed in the 1970s. It now houses a collection of model locomotives and railway memorabilia, especially local. Old locomotives and rolling-stock are displayed in the open air.

Old Royal Observatory, Greenwich, SE10. This historic building, designed by Christopher Wren (1632–1723) – himself a capable astronomer – was built in 1675 at the insistence of Charles II, primarily to develop astronomical observations making it possible to determine longitude at sea (a goal not achieved until a century later). The first Astronomer Royal was John Flamsteed (1646–1719). Now called Flamsteed House, it displays a number of old instruments. In 1884 Greenwich was adopted as the prime meridian of the world and the line of zero longitude is marked by a strip on the ground.

Royal Air Force Museum, Grahame Park Way, NW9. Lodged on a former aerodrome, this is a national aviation museum, devoted largely to aircraft that have seen service with the RAF. There are associated exhibits devoted to related topics such as propulsion units and navigation.

Royal Armouries, Tower of London, EC3. A comprehensive display of arms and armour from the 18th century onwards, particularly items related to the Tower. The indoor displays are devoted mainly to small arms – from swords to rifles – while artillery is displayed in the open air.

The Science Museum (National Museum of Science and Industry), Exhibition Road, SW7. This is not only a great national museum but one of international stature, rivalling the Deutsches Museum in Munich and the Smithsonian Institution in Washington. The range and diversity of exhibits is enormous, from scientific instruments to giant steam engines and famous locomotives and space craft. It is unusual in that it illustrates not only the history of science and technology but very recent developments such as electronic computers. In 1977 it acquired the Wellcome Museum of the History of Medicine, a remarkable collection built up by the wealthy drug manufacturer Henry Wellcome (1853–1936).

Wimbledon Windmill Museum, Windmill Road, SW19. This mill was built in 1817 and was in use until the 1870s. In 1976, after extensive restoration, it was opened as a museum devoted to the history of windmills: the exhibits include models, machinery and tools.

To these traditional museums can appropriately be added three devoted to living exhibits.

London Zoo, Regent's Park, NW1. The London Zoo has been a popular resort for Londoners and visitors since it was opened in 1828. It ranks as one of the world's great zoological gardens, with an animal population in excess of ten thousand, representing some six hundred species, from very small to very large. It is also an important research centre and is concerned with the preservation of endangered species.

Royal Botanic Gardens, Kew, TW9. Now the world's most famous botanic garden, Kew was established in a modest way in 1769 by Princess Augusta, mother of George III. After her death in 1772, he amalgamated it with his own much more considerable estate. It was much enriched by the importation of exotic plants at the instigation of Sir Joseph Banks (1743–1820), President of the Royal Society. They were opened to the public in the 1840s and now extend to about 300 acres. Banks' policy has been continued and the plants now established have been drawn from all parts of the world. Large areas are under glass, including Decimus Burton's famous Palm House. Kew is an important research centre, concerned with the identification and classification of plants (taxonomy), economic botany, the conservation of seeds and the maintenance of a large herbarium.

Chelsea Physic Garden, Royal Hospital Road, SW3. This was established in 1673 by the Worshipful Society of Apothecaries to grow plants of medical interest and this is still the basis of the cultivated area. Among those who studied botany there was Sir Hans Sloane (1660-1753) who – in his later and much wealthier years – re-established its fortunes. His (replica) statue stands in the garden: the original was removed in 1985 to the British Museum.

Some Learned Societies

It is a natural reaction for people with interests in common – whether it be gardening or music - to organize themselves into societies, and the foundation of the Royal Society was a typical example of this. Unlike many, which have withered and died, it has gone from strength to strength. Born in the very dawn of modern science, its members dabbled effectively in all its branches: not many years earlier, Francis Bacon (1561–1626) could contemplate preparing single-handedly a systematic compilation of all human knowledge so that it might be usefully applied for the benefit of mankind. As the scope of science grew, however, some compartmentalization became necessary and specialized societies began to be formed. Some were in the provinces, where science flourished in the new towns of the Industrial Revolution: others were founded in London and some of these achieved national stature. One of the earliest was the Linnaean Society, founded in 1788 to study and promote all branches of natural history. Its purpose was to commemorate the great Swedish botanist Carl Linnaeus (1707–78), whose library and collection it housed and later purchased: since the 1850s the Society has been lodged in Burlington House, Piccadilly, W1. It was there, in July 1858, that Charles Darwin and Alfred Russel Wallace jointly made public their views on organic evolution by natural selection.

Meanwhile, the Royal Society seemed to many to have become too élitist and detached from normal life. This led, in 1831, to the foundation of the British Association for the Advancement of Science (BAAS) by a group of dissidents particularly identified with Charles Babbage ((1792–1871), a zealous reformer. It held its first meeting in YORK and thereafter had a peripatetic existence, holding meetings each year in different cities in Britain, and occasionally in the Commonwealth. It appealed directly to the public and it was not uncommon for several thousand people to attend. After the Second World War it went through a difficult phase but now once again flourishes, mainly through placing emphasis on the social implications of science. Its headquarters in London are at Fortress House, 23 Savile Row, W1.

Babbage, one of Newton's successors as Lucasian Professor of Mathematics at CAMBRIDGE, was not content with the formation of the BAAS and despite having accepted Fellowship of the Royal Society in 1816 continued pressure for its reform. It had undeniably become something of a social club, many of the Fellows having no scientific attainments: this was put to an end by a revision of the statutes in 1847, though naturally this took time to work through the system. Nevertheless, by the end of the century, the Society was almost wholly scientific in both membership and management. Not until 1945, however, were women admitted to Fellowship. A new problem arose in the 1960s when, under the energetic presidency of Lord Florey

(1898–1968), a strong bid was made to admit more Fellows in the field of the applied sciences. After a rather half-hearted acceptance, this liberalizing concept was, unfortunately, dropped.

As living beings are conveniently divided into the plant and animal kingdoms, so science – until the advent of biochemistry, molecular biology and other interdisciplinary studies – fell naturally into the physical and mathematical sciences and the biological sciences. The Royal Society acknowledged this in 1887, after nearly two centuries, by dividing its *Philosophical Transactions* into Series A and B. Long before this, however, the chemical fraternity in London had decided that they needed their own professional body. This led to the foundation in 1841 of the Chemical Society, with Thomas Graham (1805–69), Professor of Chemistry at the then recently founded University College, as first President. It shortly moved to premises in Burlington House, Piccadilly, W1, where it still is, in much larger premises. In 1980 it was reorganized as the Royal Society of Chemistry.

Close by the Royal Society of Chemistry is the Royal Astronomical Society, founded in 1820. Again, one of its most energetic co-founders was Charles Babbage: its first President was Sir William Herschel (1738–1822), discoverer of Uranus.

While many of the great engineers of the 18th century were elected to Fellowship of the Royal Society, this was never really the proper milieu for them. They were for the most part practical men of affairs, often away for considerable periods organizing and supervising great enterprises. Their interest in pure science was considerable – as witness the Lunar Society of BIRMINGHAM – but it was tempered with a desire to see it put to practical use. A society dedicated to this was the Society for the Encouragement of Arts, Manufacture, and Commerce, founded in 1754. Since 1774 it has been housed (since 1908 as the Royal Society of Arts) in a splendid building at 8 John Adam Street, WC2. It played an important part in organizing the Great Exhibition of 1851 and gained great prestige from the presidency of Prince Albert 1843–61. A century later it helped to organize the Festival of Britain, when Britain – her economy severely weakened by the war – once again put her resources on public display. Over the years the Society has somewhat changed its role. Originally it encouraged industry by offering prizes for innovations, especially in agriculture. Today it still encourages innovation but mainly through lectures and fellowships.

While the Royal Society of Arts encouraged industry in a general sort of way, it did not serve the professional needs of practising engineers, who set up their own organizations. The oldest is the Institution of Civil Engineers, founded in 1818 and now located in Great George Street, SW1. It received its Royal Charter in 1828. Among its distinguished presidents have been Thomas Telford (1757–1834), John Rennie (1761–1821) and Robert Stephenson (1803–59). The Institution of Mechanical Engineers was founded in Birmingham in 1847 and moved to London 30 years later. It has been at its present premises at 1–3 Birdcage Walk, SW1, since 1899. The Institution of Electrical Engineers has twice altered its name to meet changing circumstances. In 1871 it was founded as the Society of Telegraph Engineers, to which was added 'and Electrical' in 1880. Its present title dates from 1888. A year later it moved into its present home in Savoy Place, WC2. Moves to establish a similar body for the gas industry began in 1849, but there followed unseemly wrangles between rival bodies. Finally, the present Incorporated Institution of Gas Engineers emerged in

1903. By that time the coal-gas industry had more or less reached its peak – and was beginning to experience increasing competition from electricity – but the Institution gained a new lease of life from 1966 to 1977 when Britain switched over entirely to natural gas from the North Sea.

Finally, there is the newly created Royal Academy of Engineering, at 29 Great Peter Street, SW1. This was founded in 1976, on the initiative of the Duke of Edinburgh and a group of leading engineers. It was in response to a widespread feeling within the profession that there should be a prestigious multi-disciplinary body to represent and further the understanding and professional knowledge of engineers over a wide spectrum.

LONDON COLNEY, Herts.

The de Havilland Mosquito, with the German Junkers Ju 88, was the most versatile aeroplane of the Second World War, serving variously as bomber, fighter-bomber, night fighter and transport. The prototype (1940) and two others are preserved at the Mosquito Aircraft Museum, established in 1958. Many other de Havilland types are also displayed.

LONDONDERRY, Londy

The Foyle Valley Railway Centre is devoted to the narrow-gauge systems of the Co. Donegal Railway and the Londonderry & Lough Swilly Railway. Two 1907 locomotives are on view.

LONGBENTON, Northld

Birthplace of Thomas Addison (1793–1860), founder of endocrinology. He was the first to describe the symptoms of adrenal insufficiency (Addison's disease).

LONGTON UPON TERN, Shrops.

In 1795 Thomas Telford (1757–1834) was appointed chief engineer to the Shrewsbury Canal, on which an aqueduct was needed to cross the River Tern. With the local ironmaster William Reynolds (1758–1803), Telford abandoned the conventional masonry construction in favour of a cast-iron trough. The first of its kind, it still stands although the canal has fallen into disuse.

LOUGHBOROUGH, Leics.

The Bell Foundry Museum has many exhibits relating to bellfounding from early times. The Great Central Museum contains many railway memorabilia, relating particularly to the old London and North Eastern Railway and the Grand Central. It includes an unusual collection of photographs showing successive stages of the building of the GCR in the 1890s.

Nearby Dishley Grange was the home of Robert Bakewell (1725–95), famous for his success in improving livestock – horses, cattle and sheep. 'Twopounder', his improved Leicester ram, once earned him the then remarkable fee of 800 guineas for a single season. Sadly, he dissipated his great wealth in lavish entertainment of great landowners and died a poor man.

Birthplace of Walter Weldon FRS (1832–85) who began life as a journalist – with his wife he founded the successful fashion magazine *Weldon's Journal* – but later had a successful career in the chemical industry. Weldon's process for the recovery of valuable manganese revolutionized the fortunes of the soda industry and thus also of the dependent textile industry.

LOUGH GUR, Lim.

This is the site of one of the finest stone circles in Ireland. Other well-defined circles are in the vicinity.

LUTON, Beds.

The Stockwood Craft Museum houses the Mossman Collection of more than fifty historic horse-drawn vehicles in fine condition. Additionally there are a few replicas. A separate building houses a small collection of bicycles, lawnmowers and cars.

LUTTERWORTH, Leics.

Stanford Hall Motorcycle Museum (off B5414 near Swinford) displays a large collection of vehicles in running order.

LYME REGIS, Dors.

The local cliffs have been a magnet for fossil hunters since a local girl, Mary Anning (1799–1847) discovered in 1811 a complete skeleton of icthyosaurus – which she sold to the British Museum for £25 – and subsequently plesiosaurus and pterodactylus. She became a great friend of the geologist William Buckland (1784–1856). There is a memorial window to her in the local church and a small museum on the site of her house.

The cliffs are dangerously unstable and rock falls frequently expose new fossils, especially ammonites.

Birthplace of Percy Carlyle Gilchrist (1851–1935) remembered for his part in the development of a revolutionary new process for smelting phosphoric iron ores. This immensely benefited the entire European iron and steel industry, making it possible to utilize millions of tons of previously useless ore. (See also BLAENAVON, MIDDLESBROUGH.) Birthplace also of John Gould (1804–81), son of a gardener at Windsor Castle, famous for his large and splendidly illustrated folio volumes depicting the birds and animals of the world. In all, he published forty-one volumes with 2999 plates. His assistant, the German animal painter Josef Wolf (1820–99), described him as 'the most uncouth man I ever knew'.

LYNMOUTH, Devon

The Rock Railway, working on the water-balance principle, was built in 1890 and takes passengers up the 900-ft cliff.

M

•••

MACCLESFIELD, Ches.

The town has been a centre of the silk industry since the 18th century. This is reflected in the Paradise Mill Silk Museum which displays many silk hand looms and other textile machinery, and stages demonstrations of weaving.

Birthplace of Sir James Chadwick (1891–1974), awarded a Nobel Prize for physics in 1935 for his discovery of the neutron. (See CAMBRIDGE, LIVERPOOL.)

MACHYNLLETH, Powys

Ancient capital of Wales. The Centre for Alternative Technology (3 miles N on A487) displays a wide range of sustainable technologies: wind, solar and water power, an energy-saving house, and a cliff railway.

At Eglwys Fach (6 miles S on A487) is Dyfi Furnace, a restored charcoal-burning blast furnace and a large operational water-wheel.

MAGHERAGALL, Ant.

Birthplace of the mathematical physicist Sir Joseph Larmor (1857–1942). After graduating from Queen's University, Belfast, he was for a time Professor of Natural Philosophy at Queen's College, GALWAY, before moving to CAMBRIDGE, where he was Lucasian Professor of Mathematics, a position once held by Newton, 1903–32. He bridged the gap between the old classical physics and the new physics of relativity and quantum mechanics. He did original research on the electrodynamics of electrons.

MAGHERAGHANRUSH, Sligo

Some hundreds of court graves have been located in Ireland, of which around fifty have been excavated. They are the earliest type of megalithic tomb. The mounds incorporated one or two open courts, with short burial chambers opening off them. This example stands on a hill above Lough Gill.

MAIDEN CASTLE, Dors.

Huge hillfort dating from Neolithic times but much strengthened during the Iron Age. The outer ramparts enclose 18 hectares.

MAIDENHEAD, Berks.

The brick bridge across the River Thames was built in 1838 by I.K. Brunel (1769–1849) for the Great Western Railway and is noteworthy for its extremely shallow arch. At the time, disaster was predicted but it still carries heavy traffic.

MAIDSTONE, Kent

The Tyrwhitt-Drake Museum of Carriages is one of the finest of its kind in Britain.

MALMESBURY, Wilts.

Birthplace of Thomas Hobbes (1588–1679), political and natural philosopher, noted for his work on optics and geometry. He was employed for much of his life at Chatsworth, as tutor and secretary, by the Cavendish family (see BAKEWELL, AULT HUCKNALL). It is said that having seen for the first time a proof of Pythagoras' Theorem he exclaimed: 'By God! This is impossible' and then proceeded to 'fall in love with geometry'. He abandoned the concept of light being corpuscular in nature in favour of one in which it is propagated in a medium between the source and the eye. His *Leviathan* (1651), in which he questioned the authenticity of the Bible, inevitably landed him in deep trouble: at one point a committee of bishops recommended that he be burned alive for heresy.

MALVERN, H & W

Home of the Royal Signals and Radar Establishment, where much of the development research on radar was done (see ORFORD). This research is illustrated in an exhibit in the Malvern Museum (Abbey Gateway).

The grave of Peter Mark Roget FRS (1779–1869) is in the churchyard of St James, West Malvern. Known to most people only for his famous *Thesaurus of English Words and Phrases*, which has gone through innumerable editions – twenty-eight within his lifetime – since it first appeared in 1852, he also has claims to fame as a physician and scientist. His doctoral thesis at Edinburgh (1798) was on the laws of chemical affinity – inspired no doubt by the writings of contemporary French chemists. After numerous medical appointments he became the first Fullerian Professor at the Royal Institution (1833–6) and Secretary of the Royal Society (1827–49). He invented many ingenious scientific devices, including an improved slide-rule.

Coldwall, at Herefordshire Beacon, is one of the finest examples of a British contour fort.

MANACCAN, Corn. See CREED

MANCHESTER, GM

Traditionally, but perhaps not strictly correctly, Manchester derives its name from Mancunium, a Roman settlement at the confluence of the rivers Irwell and Irk. However, the town was of little consequence until it became a centre for the wool and linen trade in the 16th century and the focal point of the cotton industry in the Industrial Revolution, earning it the name Cottonopolis until the pattern of industry changed in the 1950s. Today, it is the county town of Greater Manchester, a sprawling conurbation subsuming ten adjacent boroughs (see ASHTON-UNDER-LYNE, BOLTON, OLDHAM, SALFORD, WIGAN).

Historically, Manchester's prosperity depended more on trade than manufacture and it was this that prompted one of the great civil engineering projects of Victorian England, the 35-mile Manchester Ship Canal. From 1720 the improvement of the

Mersey and Irwell navigation made it possible to bring barges into the heart of Manchester. In 1767 the Duke of Bridgewater's Worsley Canal was extended to Runcorn, a total of 28 miles, by James Brindley (1710–72), providing an alternative route to the sea. The old Packet House where canal passengers alighted still stands beside the original dry dock. Even so, only barges – carrying no more than around 50 tons each – could be used and soon proved inadequate for the huge volume of goods. The opening of the Liverpool and Manchester Railway in 1830 – a memorable event in the history of steam transport – offered less relief than had been hoped for. Railway charges to Liverpool, and dock charges there, were ruinous. In the 1880s a ton of goods could be carried from Manchester to Calcutta for 19 shillings, but of this 12 shillings had been incurred before they ever left Liverpool. To resolve the problem a bold decision was taken: Manchester should become an inland port joined to the sea by a ship-canal capable of carrying ocean-going vessels. In the event, it turned Manchester into the fourth largest port in Britain and marked a turning-point in the city's fortunes.

The start of this great venture can be very precisely defined: it began on 27 June 1882 at The Towers, Didbury, home of Daniel Adamson (1820–90), a local ironfounder, who had covened a meeting of local businessmen to discuss the possibility of a canal. The Towers no longer exists, but its site is now marked – not inappropriately – by the Shirley Institute, an independent research laboratory for the textile trade which the canal was designed to promote.

Predictably, the proposal was violently opposed by powerful vested interests, notably Liverpool Docks and the L & M Railway. Two bills were thrown out by Parliament before one was finally approved in 1885. Then there were difficulties raising finance and the construction was dogged by ill-luck, particularly two disastrous floods in 1890. It was a massive project in terms of earth removal alone, for the canal was 35 miles long with a minimum depth of 28 ft and a width of 120 ft. Four sets of locks were needed to gain the necessary lift of 60 ft between the tidal part of the canal and the docks at Manchester. Its most remarkable feature is the unique 250-ft swing aqueduct at Barton, replacing Brindley's original masonry aqueduct across the River Irwell. The swing aqueduct is pivoted at its centre so that it can be turned to allow passage of ships: when full it holds 600 tons of water. Manchester is the centre of a complex canal system and Castlefield – Britain's first urban heritage park – is a convenient point to sample it. Here can be seen part of the old Bridgewater Canal and the beginning of the 30-mile Rochdale Canal, begun in 1794, which crosses the Pennines by way of 92 locks. The last of these is by 'Dukes 92' public house. At Portland Basin, Tameside, three of the old canals converge: the Ashton Canal, the Peak Forest Canal and the Huddersfield Narrow Canal. Also at Castlefield can be seen Victorian railway viaducts which now carry the new Metrolink rapid transport system.

The development of the railway system in Britain owes much to George Bradshaw (1801–53) who set up in business as a map engraver in Market Place. He later turned to publishing and his name is famous for the national railway timetable which first appeared, as *Bradshaw's Railway Companion*, in 1838: it was soon immensely successful as *Bradshaw's Monthly Guide*. The Quaker politician John Bright once asserted that the books he most frequently consulted were the Bible and Bradshaw. The guide was doubly helpful to travellers: not only did it give them up-to-date details of railway schedules but it brought pressure on the operators to make their trains run to time.

By many, Manchester is regarded as primarily a great commercial and industrial centre but it has long been also an important cultural centre, especially in science. The immensely learned John Dee (1527–1608), alchemist and occultist, was Warden of the Collegiate Church 1596–1608. To an extent this resulted from force of circumstances, for the Act of Uniformity of 1662 – establishing a prescribed standard of religious belief – excluded Dissenters from Oxford and Cambridge, the great national centres of learning. Accordingly, education and intellectual activities came to be organized on a local basis, particularly associated with Unitarianism and the Society of Friends. In the Manchester area this found expression in the famous, but short-lived, Warrington Academy (1757–83) where the chemist Joseph Priestley (1733–1804) taught for some years before going on to LEEDS. This was followed by the Manchester Academy (1786) where John Dalton (1766–1844) was employed, briefly, as a tutor. It had a somewhat chequered history: it moved to York in 1803, back to Manchester, and finally to OXFORD as what is now Manchester College. In BIRMINGHAM the need for a forum to discuss the scientific and industrial advances of the day was met by the foundation of the Lunar Society, which spanned roughly the second half of the 18th century. In Manchester the corresponding body was the Manchester Literary and Philosophical Society, founded in 1781 and still surviving. One reason for this is that it had from the beginning an organized constitution with formally elected officers: it was thus essentially self-perpetuating. Originally it met at Cross Street Chapel – badly damaged during an air raid in the last war – but in 1799 a house was bought at 36 George Street. It was here that Dalton – a member for 50 years and President from 1819 until his death in 1844 – had his laboratory: he lived close by at 35 Faulkner Street. Here, in 1803 he announced – quite casually – his revolutionary discovery that the atoms of different elements are distinguished by differences in their weights. This brought him international recognition, yet to the end he remained a simple-living member of the Quaker faith. On his death forty thousand people filed past his coffin and there were a hundred carriages in the funeral procession. A new street, joining Bridge Street and Princess Street, was named after him.

From boyhood until his death Dalton kept a daily record of the local weather, believing in 'the advantages that might accrue to the husbandman, the mariner, and to mankind in general, if we were able to predict the state of the weather with tolerable precision'. His *Meteorological Observations* (1793) described how to make simple meteorological instruments and commented on the aurora borealis and other atmospheric phenomena. Dalton suffered from colour blindness – his particular defect is still often known as Daltonism – and his first paper to the Lit. & Phil. was a carefully considered account of the condition based on his own experience.

A much lesser, but nevertheless interesting, chemist was Thomas Henry FRS (1734–1816), a founder member of the Society who lived in Manchester from 1764. He pioneered the use of chlorine in bleaching textiles and became known as 'Magnesia' Henry because of his promotion of sales of magnesia for medical purposes. The business sustained his family until 1935.

Another famous member of the Society was James Prescott Joule (1818–89), celebrated for his precise experimental determination in the 1840s of the mechanical equivalent of heat: this, too, was a development that revolutionized scientific thought. Joule gave the first public account of his work in 1847 in the reading room of St Ann's Church in St Ann's Square. For this reason it is often referred to as the St Ann's Church Lecture.

Statues of Dalton and Joule stand in the Town Hall, which also has twelve large murals by Ford Madox Brown, which took him 15 years to complete. One depicts Dalton collecting marsh gas from a pond, closely watched by a group of children. Another shows William Crabtree (1610–44) of Broughton – a merchant with a comprehensive knowledge of astronomy – observing a transit of Venus in 1639, and thereby hangs a curious tale. Crabtree had been befriended by another keen amateur astronomer Jeremiah Horrocks (1618–41) who lived at Liverpool. In a sadly brief life Horrocks turned his attention to virtually every aspect of contemporary astronomy, but in particular to correcting Kepler's famous Rudolphine tables (1627), which set out in detail the laws of planetary motion. Among other things, they made it possible to predict far in advance such notable events as solar and lunar eclipses. They also made it possible to predict the rare transits of Venus, in which the planet visibly passes between the Earth and the Sun. It was to observe the next transit, in 1769, that the Admiralty dispatched Captain James Cook to Hawaii in *Endeavour* (see WHITBY). Horrocks detected an error in Kepler's Tables and calculated that – unsuspected by any other astronomer in the world – a transit would occur on 4 December 1639. He and Crabtree independently observed it – for the first time in the history of astronomy – and Horrocks' observation made it possible to make important corrections to accepted constants for the orbit of Venus. Sadly, Horrocks died at the age of 23 and gained no recognition in his lifetime but his papers were published posthumously, many by the Royal Society. Crabtree, too, made important corrections to the Tables but little is known of him after Horrocks' death. Ford Madox Brown's fresco shows a curious quirk. He had virtually nothing to guide him as to Crabtree's appearance and so gave him the head of his own contemporary C.B. Cayley (1823–83), famed for his scholarly translations of classical literature.

In 1940, 36 George Street was almost completely destroyed in an air raid and almost all its Dalton and Joule memorabilia and other historic treasures were destroyed. A new building has since been erected on the same site.

Another member of the Society, though more industrialist than scientist, was Robert Owen (1771–1858) an important innovator in the cotton industry and a great social reformer (see NEW LANARK). From 1813 until his death Manchester was the home of the great Victorian engineer William Fairbairn (1789–1874). Self-taught, he set up an engineering works in 1817, at what is now the Arndale Centre, High Street. His interests were very diverse: riveting machines, water-wheels, iron ships, and bridges. At the end of his life he claimed that he had built nearly a thousand bridges. Of these the most memorable was the tubular bridge over the Menai Straits at BANGOR, built with his close friend Robert Stephenson (1803–59).

Yet another Victorian engineer who achieved fame in Manchester was Joseph Whitworth (1803–87) who established his first modest workshop in 1833 in Chorlton Street. In 1880 a new and much bigger works was built at Openshaw. Whitworth is particularly remembered for his high-precision work. By 1855, he could measure to an accuracy of one-millionth of an inch while many of his contemporaries were still content with 'a bare sixteenth' or 'a full thirty-second'. He also pioneered the standardization of screw threads.

Another self-taught man who is remembered for his scientific work is William Sturgeon (1783–1850), born at Whittington near Kirkby Lonsdale. He began life as an apprentice cobbler and spent nearly twenty years as a private in the Royal Artillery. He contrived to teach himself Latin, Greek and natural science, and took a particular interest in the then little-understood science of electricity. In 1840 he was appointed Superintendent of the Victoria Gallery of Practical Science in Manchester, a short-lived institution founded to 'stimulate research and foster inventive talent': briefly, one of his students was Joule. This failed in 1842 but he eventually received a very small state pension. Sturgeon was a practical inventor rather than a theoretician. He made a practical electric motor, introduced the soft-iron electromagnet, and invented the commutator. Historians of science now acknowledge the importance of his work, but he gained little recognition in his lifetime. There is a memorial tablet in Kirkby Lonsdale church. Ironically, this was destined for Prestwich church, where he is buried, but the vicar there thought him of too little account to be so honoured. Among Sturgeon's friends was John Dancer (1812–87) who came to Manchester in 1841 and established a successful business as a scientific instrument maker and optician at 43 Cross Street. He was the inventor of microphotography and began to exploit it commercially in 1857. For a decade microphotographs were a fashionable craze. Queen Victoria had a signet ring carrying a family portrait only one-eighth of an inch in diameter: it was magnified by a jewel cut as a lens. Contemporary scientists did not take him seriously but today microfilm is very widely used, for example in microfiche reproduction of books where thin photographic film only 6 in. x 4 in. can carry nearly a hundred pages of printed text. He also invented a stereoscopic camera.

The Lit. & Phil. was by no means the only manifestation of organized scientific activity in 19th-century Manchester. John Owens (1790–1846) made a large fortune as a textile merchant and railway speculator: on his death he left £96,000 to endow a new university. Owens College was founded in 1851 in the house of the political reformer Richard Cobden (1804–65) and received its Parliamentary charter 20 years later. For a time it was associated with University College, LIVERPOOL, and Yorkshire College, LEEDS, but at the turn of the century all three gained independent university status. The new foundation had from the outset an exceptionally strong department of chemistry. The first professor was Sir Edward Frankland (1825–99), remembered for the concept of valency as determining the constitution of molecules. He was followed in 1857 by Sir Henry Roscoe (1833–1915), a versatile research chemist and a dedicated teacher. He was a Member of Parliament for South Manchester 1885–95 and his active public life did much to further the interests of the new foundation, which grew to possess what was indisputably the leading chemistry school in Britain. He was succeeded in 1892 by W.H. Perkin Jr (1860–1929), son of Sir William Perkin (1838–1907) whose synthesis of mauve in 1856 was the foundation of the modern dyestuffs industry. He was a brilliant research worker, with a particular interest in the chemistry of natural products and, like Roscoe, a gifted teacher.

Among Perkin's students, and later his collaborator, was Robert Robinson (1886–1975). In 1912 he left to take up a professorship at Sydney and then came back to occupy chairs successively at LIVERPOOL and ST ANDREWS before returning to Manchester as professor in 1922. Meanwhile Perkin had gone to OXFORD to take up the prestigious Waynflete Professorship, and in due course Robinson succeeded him there in 1929. Robinson was perhaps the last, and greatest, organic chemist in the classical tradition, depending on simple apparatus, a keenly analytical mind and marvellous intuition. He too rose to fame: Nobel Laureate (1947), President of the Royal Society, Order of Merit.

For some years Robinson worked in Perkin's private laboratory and there, one day in 1904, an unexpected foreign visitor announced himself. This was Chaim Weizmann (1874–1952), an ardent Zionist destined to be the first President of the new state of Israel. At that time, however, he was simply a well-qualified organic chemist looking for a job. Perkin was impressed and engaged him first as a research assistant and later as lecturer in biochemistry. In 1912 he discovered a bacterium (*Clostridium acetobutylicum*) which broke down starch into ethanol, acetone and butanol. During the First World War there was a desperate shortage of acetone for plasticizing cordite, and Weizmann organized its large-scale production by fermentation for the Ministry of Munitions. He sought no reward for himself but after the war secured from Lord Balfour the famous Balfour Declaration, a promise of support by Britain in establishing a national home for the Jews. The Jewish Museum, 190 Cheetham Hill, is a memorial to him.

As industrial activity increased and became more complex there was a demand for better-trained industrial workers. In the first half of the 19th century the majority of these were illiterate: in 1841 one-third of the men and half the women marrying in England and Wales signed the register with a cross. Despite, or perhaps because of this, there arose in the 1820s one of the great movements in British educational history – the mechanics' institutes, which had a powerful influence up to the Technical Instruction Act of 1889. Many of these were the parent bodies of nationally important institutions. One such was the Mechanics' Institute of Manchester, founded in 1824, which by a complicated process of evolution is now the independent University of Manchester Institute of Science and Technology (UMIST) in Sackville Street, with a graduate research staff of over a thousand.

It is said that the chemical industry grew up in the shadow of the textile industry, so it is not surprising that science in Manchester should always have had strong chemical connotations. Nevertheless, achievements in physics, too, have been impressive, as exemplified by the work of Joule. Physicists like Joule, who for many years collaborated with Kelvin at GLASGOW, bridged the gap between the old classical physics of Newton and the revolutionary new physics of the 20th century, with its emphasis on atomic and sub-atomic structure. Here, too, Manchester has played an outstanding role.

The acknowledged founder of this new branch of science was Ernest Rutherford (Lord Rutherford of Nelson, 1871–1937). A New Zealander by birth, he did much of his early research on radioactivity while Professor of Physics at McGill University, Canada, and concluded a remarkable career at the Cavendish Laboratory in CAMBRIDGE. But it was during his time as Professor of Physics at Manchester University 1907–19 that his most original experimental and theoretical work was done. It was there that he formulated in 1911 the crucially important concept of the atom consisting of a relatively heavy positively charged nucleus surrounded by small negatively charged electrons orbiting round it. Though fundamentally correct, classical electrodynamics could not explain why the electrons did not spin down and fuse with the nucleus. The difficulty was solved in the following year when a brilliant young research worker, Niels Bohr (1885–1962), came from Copenhagen to work with him for three months. Bohr recognized that Max Planck's new Quantum Theory, enunciated at the turn of the century, required the electrons to move in fixed orbits, transferring from one to another only when radiation was emitted or absorbed. Later, Bohr returned to Manchester as Reader in Theoretical Physics 1914–16. The Rutherford/Bohr atom can fairly be described as the foundation of atomic physics and Manchester as its birthplace. There is a commemorative plaque in Oxford Street, at the corner of Coupland Street. Rutherford won a Nobel Prize in 1908 and went on to be President of the Royal Society, elevated to the peerage and appointed to the Order of Merit. Bohr, too, won a Nobel Prize, in 1922, but by then he was back in Copenhagen.

Rutherford's departure for Cambridge did not signal any decline in physics at Manchester. His successor was Sir Lawrence Bragg (1890–1971) who – with his father Sir William Bragg (1862–1942) – had already made his name as a pioneer of the use of X-rays to determine crystal structure. For this they had shared a Nobel Prize in 1915. In due course he was Director of the National Physical Laboratory, succeeded Rutherford at Cambridge, and finally was Director of the Royal Institution in LONDON. But even then the wave had not broken. Bragg was followed by Patrick Blackett (Lord Blackett of Chelsea, 1897–1974) who changed the direction, but not the quality of research by building up a group to study cosmic rays. In 1935, however, he was appointed to the Air Defence Committee and from then until the end of the Second World War he was much preoccupied with problems of military technology and strategy. Back in Manchester in 1945, he encouraged his colleagues to look not only at the micro-world of atomic particles but outward to the vast expanses of the Universe. In particular, he encouraged Sir Bernard Lovell in his pursuit of the new science of radio astronomy, studying the structure of the universe not from the visible light we receive from it, as in traditional astronomy over the centuries, but through the much longer waves emitted by 'radio stars'. Conspicuous evidence of this is to be seen at JODRELL BANK, where the world's first giant radio telescope is a landmark for miles around. Completed in 1957, it has a steerable bowl 250 ft in diameter. Blackett's personal interest turned to terrestrial magnetism and in 1953 he moved to Imperial College, London, as Professor of Physics. Like his predecessors, Blackett earned high honours. These included a Nobel Prize (1948), Presidency of the Royal Society and the Order of Merit.

Apart from the commemorative plaques already mentioned, a number of others are worth noting. The one at the junction of Corporation Street and Balloon Street marks the first manned balloon ascent in Manchester (12 May 1785), by the pioneer aeronaut and engineer James Sadler (1753–1828), fresh from similar exploits in OXFORD. That at 70 Bridge Street marks the site of the first Provincial School of Anatomy, founded in 1814 by Joseph Jordan (1787–1823). Henry Royce (1863–1933) opened an engineering workshop in Cooke Street in 1884 to manufacture electrical equipment: he built his first motor-car in 1904 and soon began to produce the famous Rolls Royce series in partnership with C.S. Rolls (1877–1910). St Mary's Parsonage marks the site of the world's first municipal gasworks (1824), built surprisingly, by the local police commissioners: it grew from a very small plant built to light a lamp over the police station door.

At the turn of the century it was not only the advent of the motor-car that heralded the rise of a vast new industry. The Wright Brothers' historic flight in 1903 also presaged a huge new industry – the manufacture of aircraft. Here, too, Manchester was early in the field. Alliott Verdon-Roe (1877–1958), born at Patricroft, served as an engineer in the merchant navy 1899–1901 and became keenly interested in the flight of birds, especially the albatross, and this led him to devote himself to the problems of mechanical flight. Within weeks of their flight he was in touch with the Wright brothers and by 1908 had himself flown in an aeroplane of his own design. It was the first time that an aeroplane designed and built in Britain had taken off under its own power. In 1910 he founded A.V. Roe Limited (Avro) in Manchester, initially at Brownsfield Mills and later at Miles Platting. By 1911 he had built the world's first aeroplane with totally enclosed cabin, followed in 1913 by the extraordinarily successful Avro 504 biplane. During the First World War, Avro built ten thousand of these aircraft and an equal number were built by other firms under licence. Remarkably, it remained the standard trainer aircraft until 1939.

His first flight was made at Brooklands, later to become an important centre of aircraft design and construction, and in 1954 a memorial plaque was erected there to ' . . . the first of a long line of famous pioneers and pilots . . . on this flying field of Brooklands'. In the same context two Mancunians were the first airmen, in 1919, to achieve a non-stop flight across the Atlantic. They were John William Alcock (1892–1919) and Arthur Whitten Brown (1886–1948). There are memorial plaques at 6 Kingswood Road, Fallowfield, and 6 Oswald Road, Chorlton-Cum-Hardy, respectively. A sculpture by Elisabeth Frink at Manchester Airport commemorates the flight and there is also a commemorative statue at Heathrow Airport, London.

MARGATE, Kent

Draper's Windmill in College Road, built in 1847, has been restored to working order.

MARLBOROUGH, Wilts.

Birthplace of Thomas Hancock (1786–1865), pioneer of the industrial use of rubber. Particularly important was his process for the vulcanization of rubber with sulphur, patented in 1843. This gave a hard durable material with many uses.

MARSHFIELD, Wilts.

The history of agriculture and dairying, both local and general, is depicted in the Castle Farm Museum. It is housed in a farmhouse and outbuildings dating from the late 18th century, but includes also a fine long-house dating from the second half of the 16th century.

MARVAL, Corn.

Birthplace of John Mayow FRS (1641–1679), physician noted for his experiments on respiration. He clearly recognized the chemical similarity between respiration and combustion.

MARYPORT, Cumb.

Maryport, now a quiet seaside town, was once a thriving port active in shipbuilding, sail-making and related industries. The small Maryport Maritime Museum depicts these once-important local activities.

MATHRY, Dyfed

The Long House (Samson's Stone) is a Neolithic chambered tomb from which the soil has disappeared. Seven massive uprights support a huge capstone.

MATLOCK, Derbs.

The National Tramway Museum at Crich houses a unique collection of restored horse, steam and electric trams. The Peak District Mining Museum is at Matlock Bath, in the Pump Room where visitors used to take the waters. It is devoted primarily to local lead mining, which has been practised since Roman times. Its most striking exhibit is a giant water-pressure engine built by Richard Trevithick (1771–1833).

Matlock Bath Model Railway Museum is of historic interest as it uses miniature versions of early LMS locomotives and the layout mimics the original railway system at the old station and quarry sidings at Millers Dale.

MAYNOOTH, Kild.

St Patrick's College (1795) is Ireland's leading theological college for training Roman Catholic clergy. This interest is reflected in the College Museum, but this also includes a scientific section, based on a collection made by Dr Nicholas Callan (d.1864). Apart from his own batteries, induction coils and other electrical items, there have since been added a variety of later electrical apparatus, including part of Marconi's 1898 wireless equipment.

MELROSE, Bdrs

The Melrose Motor Museum displays a good collection of vintage and veteran cars, motor-cycles and bicycles. All are in running order. A large and varied collection of other items relating to motoring are also on show.

Eildon Hill (1 mile SE) is the largest hillfort in Scotland, once capital of the Selgovae.

MERTHYR TYDFIL, M Glam.

Once a great centre of the iron industry: in the mid-19th century the town had the highest concentration of ironworks in Britain. Here, in 1804, at the Pen-y-Darran Iron Works, Richard Trevithick (1771–1833) built the first steam locomotive to run on rails; it hauled a load of 10 tons and 70 men in five wagons. There is a full-size working model in the Welsh Industrial and Maritime Museum, CARDIFF. The Ynysfach Engine House Museum has many exhibits relevant to the industry. At Dowlais are the remains of a very large blowing engine. Another reminder of the industry's past history is an iron aqueduct at Pont-y-Cafnau built to carry the Glamorgan Canal over the River Taff.

The local iron industry was founded by Anthony Bacon (1718–86) who in 1765 leased land at Cyfarthfa rich in coal and minerals. Cyfarthfa Castle – a great Gothic mansion built in 1825 and now a museum – is a memorial to the ironmaster William Crawshay II (1788–1867).

Michael Faraday vividly described a visit he made to Merthyr in 1819: 'The operations were all simple enough but from their extensive nature, the noise which accompanied them, the heat, the vibration, the hum of men, the hiss of engines, the clatter of shears, the fall of masses, I was so puzzled I could not comprehend them . . . ' Faraday's host was Josiah Guest FRS, one of a dynasty of ironmasters and coal owners. By 1850, twelve thousand people depended on his huge Dowlais works. Today, the family name is remembered in a far larger combine – Guest, Keen, and Nettlefold.

MIDDLESBROUGH, N Yorks.

A port on the estuary of the River Tees founded in 1829 by a group of Quakers associated with the Stockton and Darlington Railway, it grew slowly until the 1850s, when the discovery of the main seam of Lias ironstone in the Cleveland Hills led to a rapid expansion. This was further increased in 1879 when the introduction of the Gilchrist and Thomas process made it possible to smelt the local phosphoric ores.

The Captain Cook Birthplace Museum is in Stewart Park, near the memorial marking the site of the cottage (now re-erected in Melbourne, Australia) in which Cook was born in 1728. It follows his life up to his death in 1779 in Hawaii, and the exhibits include a reconstruction of the below-deck accommodation in *Endeavour*. The Newham Grange Leisure Farm Museum (Coulby Newham) is situated in a working farm. Exhibits include agricultural machinery, and reconstructions of a vet's surgery and a saddler's workshop.

MIDDLETON, Derbs.

Arbor Low (3 miles W) comprises a large Bronze Age henge, some 80 m in diameter, and fifty stones – no longer upright – in a circle.

MIDDLE WALLOP, Hants.

The Museum of Army Flying illustrates every aspect of army flying from the 19th century: balloons, kites, airships, gliders and helicopters.

MIDHURST, W Ssx

Richard Cobden (1804–65) was born in Heyshott (3 miles S) and is buried in the churchyard of St Mary Magdalene, West Lavington (1 mile E). Although best known as an economist and politician, very actively involved in the repeal of the Corn Laws (1846), as a young man he was a successful industrialist in Manchester, where he had a share in a profitable calico-printing mill: it was this that made it possible for him to embark on a political career. Unfortunately, this brought about his financial downfall but eventually private subscription restored his fortunes and made it possible for him to purchase his old home, Durnford, at Heyshott.

MILL GREEN, Herts.

The three-storey water-mill dates from the 18th century, but was modified early in the 19th. Its machinery, now constructed in wood and iron, has been fully restored. There is an associated Old Mill House Museum.

MILLOM, Cumb.

Swinside is a fine Bronze Age stone circle dramatically situated on the open fells. It comprises fifty-five slate blocks set in a ring 27 m in diameter.

MILTON KEYNES, Bucks.

Site of the Open University, established by Royal Charter in 1969 to meet the needs of students throughout the UK.

MILVERTON, Som.

Birthplace of the versatile scientist Thomas Young (1773–1829), famous equally for establishing the wave theory of light and for his outstanding contribution to the deciphering of Egyptian hieroglyphics, beginning with his study of the Rosetta Stone.

MINCHINHAMPTON, Glos. See SHERBORNE

MINIONS, Corn.

The Hurlers are three aligned stone circles on the east side of Bodmin Moor, dating from the Bronze Age.

MONMOUTH, Gwent

In the centre of the town is a monument to the aviator C.S. Rolls (1877–1910), son of the first Baron Llangattock, of Hendre, Monmouth. With F.H. Royce (1866–1933) he founded the famous Rolls Royce motor-car manufacturing company. He was the first man to fly from England to France and back without stopping. He was killed in a flying accident in the following year, the first English victim of aviation.

MONTROSE, Tays.

Birthplace of the botanist Robert Brown (1773–1858) whom Humboldt regarded as *botanicorum facile princeps*. His most important work was as curator of Sir Joseph Banks' vast range of specimens – the core of the British Museum's natural history collection – but he is most widely remembered for his discovery of the Brownian Motion, the dancing movement of tiny particles, such as pollen, suspended in water. The Public Library, built on the site of his house, displays a bust of him.

The Sunnyside Museum is situated in Sunnyside Royal Hospital, founded as an asylum in 1781, the first in Scotland. Exhibits illustrate the history of psychiatry and the treatment of the insane in Scotland since the late 18th century.

MORETONHAMPSTEAD, Devon

Scorhill, a large Bronze Age stone circle, lies 6 miles N on Dartmoor. It originally consisted of about forty stones, of which twenty-three still stand and three have fallen.

MORPETH, Northld

Birthplace of William Turner (*c.* 1508–68), the 'Father of English Botany'. His

Herball, the first in English to include original material, was published in three parts (1551/62/68).

MORWELLHAM, Devon

Port on the River Tamar. The building of the Tavistock Canal in 1817 ushered in a period of great prosperity until the 1870s for the carriage of minerals, especially copper ore, to Plymouth, 23 miles downstream. The canal ended 70 m above river level and goods were conveyed to the quayside by an inclined plane, 230 m long, which still survives.

MOUSA, Shet.

Although one of the finest prehistoric monuments in the British Isles, the Broch of Mousa on Sandwich Island (7 miles S of Lerwick) is little-known because of difficult access. The walls still stand to a height of 13 m (cf NEWBIGGING).

MUCH HOOLE, Lancs.

Home of Jeremiah Horrocks (1618–41), astronomer. He was self-taught, died at the age of 23, and was no more than the curate of St Michael's Church, Much Hoole. Yet he made astronomical history by observing in 1639 a transit of Venus across the sun, an extremely rare event. He and his friend William Crabtree (1610–44) of MANCHESTER alone observed it because he had detected an error of two days in Kepler's famous Rudolphine Tables, compiled to predict such occurrences. Kepler's mathematical computations correctly forecast a transit far ahead in 1761 but completely overlooked the imminent one in 1639. Horrocks also observed that Kepler's estimate of the diameter of Venus was ten times too large.

Horrocks' brilliant work was posthumously brought to the attention of the scientific world by John Flamsteed, first Astronomer Royal, Isaac Newton and others. There is a commemorative sundial at the church inscribed *Sine sole sileo* (Without the Sun I am silent).

MULLINAHONE, Tipp.

Threshing is one of the most important of all agricultural operations. The Threshing Museum is devoted to all its aspects, illustrated by machinery – balers, winnowing machines, reapers and binders – mostly within the time range 1920–40.

MULLION, Corn.

Halligye (3 miles N) is a T-shaped Iron Age souterrain nearly 20 m long. It is the best-preserved in Cornwall.

N

NAAS, Kild.

The Punchertown Stone is the tallest (6 m) standing stone in the British Isles. There is another nearly as tall (5 m) close by at Forenaghts.

NAILSEA, Som.

The Nailsea Glass House was founded in 1788 and flourished for more than a century. It made mainly window glass but later extended to domestic ware, which found many imitators.

NANTGARW, S Glam.

Former mining village, famous for the lustrous products of the Nantgarw China Works, founded in 1813.

NANTWICH, Ches.

Birthplace of Sir William Bowman (1816–92), pioneer histologist and a great ophthalmic surgeon: one of the first to perform the operation of iridectomy for the treatment of glaucoma. Bowman's Capsule in the kidney is named after him. Nantwich is also the birthplace of John Gerard (1545–1612), whose great *Herball* of 1597 is a famous review of botanical knowledge and outlook at the end of the 16th century.

NARBERTH, Dyfed

Blackpool Mill was built in 1813 on the site of a disused iron forge. In recent years the mill and its machinery have been substantially restored.

NASEBY, Northants.

Naseby is famous for the Battle of Naseby (1645) which effectively decided the outcome of the Civil War. This event is commemorated in Naseby Battle and Farm Museum, which also exhibits craftsmen's tools and agricultural equipment. Additionally, there is a display of vintage British and American tractors.

NEATH, W Glam.

An important mining centre. The Cefn Coed Colliery Museum displays mining machinery and working conditions in this former mine. The beautiful Aberdulais Falls (3 miles NE) are now a great tourist attraction (NT), but their water power was for centuries used for milling corn, working iron and smelting copper, and, lastly, to supply a tinplate works. Relics of this industrial activity are much in evidence.

NEFYN, Gynd

Famous for the mineralogical phenomenon of the Singing Sands. The physical configuration and hardness of the sand grains is such that when rubbed together – as when walked upon – they emit a high-pitched sound.

NELSON, Lancs.

The Pendle Heritage Museum, Burrowford, illustrates the history of Park Hill, a 17th-century manor house, home of the Bannister family. It includes the earlier Park Hill Barn, which houses a small collection of agricultural machinery and equipment.

NETHER ALDERLEY, Ches.

The restored 15th-century corn mill has two overshot water-wheels. They are unusual in being arranged in tandem. The water is fed first to a 13-ft wheel and then to a somewhat smaller one.

NEVERN, Dyfed

Pentre Ifan is a burial chamber from which the covering mound of soil – once 40 m long – has disappeared. Three massive uprights support a huge 17-ton capstone.

NEWARK, Cambs.

Birthplace of Sir Godfrey Hounsfield (b.1919), Nobel Laureate 1979, famous for the development of computer-aided tomography (body scanning). This is probably the greatest advance in medical diagnosis since Roentgen's discovery of X-rays in 1895.

Millgate Folk Museum, in an old warehouse beside the River Trent once owned by the Trent Navigation Company, illustrates local history from Victorian times up to 1940. Exhibits include tools and equipment used in crafts such as printing, blacksmithing and malting, as well as in agriculture.

Newark Air Museum (Winthorpe) displays some forty aircraft, mostly fighter aircraft and trainers from the Second World War. One of the most impressive is an Avro Vulcan B2.

NEWBIGGING, Tays.

Many Iron Age settlements possessed souterrains, systems of long, narrow underground passages, some consisting of a revetted trench covered with stone slabs, and others tunnelled through the rock. Their function is obscure. They are sometimes supposed to have been places of refuge but as such would have been very vulnerable. Possibly they served the more mundane purpose of keeping food cool – a megalithic larder. The Ardestie souterrain is a fine example, though lacking the original roof slabs. Another, with several entrances, is located close by at Cairlungie.

NEWBRIDGE, Kild.

Birthplace of Kathleen Lonsdale (née Yardley 1903–71), famous for her pioneer research on X-ray crystallography. She was one of the first two women to be elected to

Fellowship of the Royal Society, in 1945: the other was Marjory Stephenson (1885–1948), a biochemist. Much of her research was done in LONDON, at the Royal Institution and University College.

NEWBURY, Berks.

Newbury District Museum (The Wharf) concerns itself primarily with local history, but it also displays the George Park History of Photography Collection, comprising more than sixty cameras and much related material.

NEWBY BRIDGE, Cumb.

The Industrial Revolution is closely identified with the extensive use of iron, but wood still had an important part to play. Stott Park Bobbin Mill (2 miles N) was one of many mills in the Lake District which supplied bobbins and reels for the textile industry and cotton reels for the domestic trade. Much of the 19th-century machinery and equipment has been preserved and there are occasional demonstrations.

NEWCASTLE EMLYN, Dyfed

The Museum of the Woollen Industry is situated at Dre-Fach Felindre (3 miles E). It displays a comprehensive collection of relevant equipment and memorabilia.

At Maenachlog-ddu is Gors Fawr circle of fifteen stones. It bears a remarkable resemblance to that at FERNWORTHY on Dartmoor, a hundred miles away across the Bristol Channel.

NEWCASTLE UPON TYNE, T & W

The cultural, commercial and administrative centre of north-east England, supporting much heavy industry. The Museum of Science and Technology has a wide range of exhibits, especially ones relating to engineering and shipbuilding. A major exhibit is Charles Parson's *Turbinia* – now at Newcastle Discovery, a new display of technology and Tyneside's industrial past – which at the Spithead Naval Review in 1897 dramatically demonstrated the efficiency of his newly perfected steam turbine. Many scientific instruments are also displayed. The high-level bridge across the River Tyne was built by Robert Stephenson (1803–59) in 1849.

Thomas Addison (1793–1860) was born locally at LONGBENTON. He described what is now known as Addison's disease – a defect of the adrenal glands – in 1855. Also the birthplace of Neil Bartlett (b.1932), famous for his discovery, at the University of British Columbia, of compounds of the inert gases of the atmosphere, until 1962 supposed to be incapable of any kind of chemical reaction. The geologist Arthur Holmes (1890–1965) was born at HEBBURN. He is remembered for his use of radioactive methods to determine the age of rocks.

The city can fairly be described as the birthplace also of the modern electric lamp, for it was here that Sir Joseph Swan (1828–1917) first publicly demonstrated his incandescent filament lamp on 18 December 1878 to the Newcastle upon Tyne Chemical Society. In the following year he used electricity to light his own home. In 1881 he began to manufacture lamps at Benwell, 3 miles W of Newcastle.

NEW GALLOWAY, D & G

Glenlair at Parton (10 miles SE) – a mansion set in an estate of several thousand acres – was the family home of James Clerk Maxwell (1831–79), one of the greatest of all theoretical physicists and formulator of the electromagnetic theory of light. After holding professorships at Aberdeen and King's College, London, he retired from academic life in 1865 to enlarge Glenlair (sadly, gutted by fire in 1929). There he wrote his *Treatise on Electricity and Magnetism* (1873) one of the great classics of scientific writing. He was buried in the churchyard at Parton and there is a memorial plaque at the entrance. There is also a commemorative window in the nearby church at Corsock: it depicts the Magi following the Star of Bethlehem. A Greek inscription is based on the Epistle of St James (1: 17): 'Every good gift and every perfect gift is from above, and cometh down from the Father of lights, with whom is no variableness neither shadow of turning' – an oblique scholarly reference to Maxwell's brilliant work on the electromagnetic theory of light.

NEWGRANGE, Meath

Site of one of the most spectacular passage-caves in Ireland – indeed, in the whole of Europe. The mound covers nearly an acre and contains a 60-ft passage way leading to a cruciform chamber with a corbelled roof. The massive stone outside the entrance is engraved with a triple spiral unique in Ireland. There are also well-preserved passage-graves nearby at Dowth and Knowth.

NEWINGTON, GL

Birthplace of Michael Faraday (1791–1867), famous for his research on electro-magnetism at the Royal Institution, LONDON.

NEW LANARK, Strath.

One of the great monuments of the Industrial Revolution. Here, on the Falls of Clyde just south of Lanark, Richard Arkwright (1732–92) and David Dale (1739–1806) established New Lanark, an enlightened project in which cotton mills were integrated with workers' housing. By 1800 the population had risen to 2500. In that year Robert Owen (1771–1858), the social reformer and philanthropist, arrived from NEWTOWN, Powys. Arkwright having died, he bought New Lanark from Dale and married his daughter. Under his guidance the project was further developed to include a Nursery (1809), an Institution for the Formation of Character (1816) and a School (1817). Much of this has survived and latterly has been steadily restored.

NEWMARKET, Cambs.

Devil's Dyke is a linear earthwork running across Newmarket Heath. It dates from the Dark Ages, but its purpose is obscure. Possibly it was part of a larger work built to protect the East Anglians from Mercia (cf OFFA'S DYKE, WANSDYKE).

NEW MILTON, Hants.

The Sammy Miller Museum has a fine collection of historic motor-cycles, especially ones used for trials and racing. Exhibits include a Norton of 1905 – the earliest known to survive – a 1908 Triumph, and the 1939 AJS Four, the first machine to lap a Grand Prix course at 100 mph.

NEWPORT, Gwent

The famous suspension bridge across the River Usk was built by I.K. Brunel (1806–59), engineer to the Great Western Railway, in 1852. Just a century later it was reconstructed as the world's first welded – as opposed to riveted – box-girder railway bridge.

NEWQUAY, Corn.

Trerice (3 miles S), a National Trust property, has a small, but very comprehensive, museum devoted to the development of the lawn mower over the last 150 years. There are some ninety exhibits in all.

NEWRY, Down.

Birthplace of Sir Joseph Barcroft (1872–1947), physiologist famous for his research on the respiratory function of the blood.

NEWTON GRANGE, Loth.

The Scottish Mining Museum (Lady Victoria Colliery) exhibits a steam winding-engine and other coal-mining equipment, as well as depicting life in a mining village.

NEWTOWN, Powys

Market town on the River Severn which in the late 18th century became an important centre for the production and distribution of Welsh flannel and tweed. This interest is reflected in the Newtown Textile Museum. An impressive building is Pryce Jones's Royal Welsh Warehouse, established in 1859 as a mail-order business.

Newtown's most distinguished citizen was Robert Owen (1771–1858), the industrialist and social reformer, particularly remembered for his NEW LANARK venture, a model town in which cotton mills were integrated with houses for the workers, and amenities such as a nursery and a school. The Robert Owen Memorial Museum commemorates his life.

NORHAM, Northld

The old railway station, built in 1847, has been preserved exactly as it was when closed in 1964. It includes the signal-box with its original traffic-control instruments.

NORMANTON, Leics.

Normanton Church Water Museum is unique. Since the 1970s it has been entirely surrounded by Rutland Water, the largest man-made lake in Europe. To preserve it, the outer walls of the church (1764) have had to be rendered waterproof and the internal floor level raised several feet. Appropriately, a major display is devoted to the history of water supply since Roman times.

NORTHAMPTON, Northants.

Town long famous for its boot and shoe industry. In the 1870s, two-fifths of those employed in the town were in this industry, by which time factories were beginning to displace hand craftsmen. The Museum of Leathercraft illustrates the preparation and use of leather from ancient Eygptian times.

NORTH BERWICK, Loth.

The Museum of Flight at East Fortune Airfield has a large collection of aircraft and engines.

NORTHILL, Beds.

Birthplace of Thomas Tompion (*c.* 1639–1713), horologist. Among his outstanding instruments were two clocks with 13-ft pendulums and a 2-second beat for the new Royal Observatory, GREENWICH (1676).

NORTHLEACH, Glos.

Gloucestershire has always been a rich farming area and this is reflected in the Cotswold Countryside Collection, located in the old courtroom and gaol (1791). It displays a fine collection of farm wagons and horse-drawn agricultural machinery.

NORTH LEIGH, Oxon.

Site of a large 4th-century Roman villa, of which a substantial part has been excavated. This has exposed fine mosaics and a bath-house.

NORTHWICH, Ches.

The ready availability of limestone, coal and salt was a main reason for north-west England becoming a major centre of the chemical industry. The Salt Museum displays salt-making equipment and other exhibits relevant to the industry.

As elsewhere industry depended heavily on canals for transport. The Anderton Lift, joining the River Weaver with the Trent and Mersey Canal, completed in 1875, was one of the great achievements of the canal-building age.

NORWICH, Nflk

Formerly one of the great centres of the woollen industry – the fine fabric known as worsted derives its name from Worstead (20 miles N). Today, its best-known industry is mustard, originated by Jeremiah Colman in the 18th century at Stoke Holy Cross. In 1854 his great nephew J.J. Colman (1830–98) moved the business to Carrow, on the outskirts of the city. It prospered and expanded to include many ancillary activities – a sawmill, cooperage, tin shop, printing works and much else. Colman's became a household name throughout the Empire. The story is told in Colman's Mustard Museum in Bridewell Alley.

At Strumpshaw (8 miles E), the Strumpshaw Hall Steam Museum displays steam-powered and other agricultural machinery. The Bridewell Museum of Norwich Trades and Industries features local clocks and equipment related to the textile, shoemaking, and other local industries. The John Jarrold Printing Museum displays printing presses, bookbinding equipment, and other related memorabilia.

The physician John Caius (1510–73), co-founder of Gonville and Caius College, Cambridge, was born here and for a time was a medical practitioner in the town. He was the first in England to teach practical anatomy, at the Barber-Surgeons Company, London. Sir Thomas Browne (1605–82) was also for many years a leading physician in the city. He was also a distinguished scholar and writer, whose works are still widely read. His *Religio Medici*, in which he asserted his right to his own opinions where these were not specifically contrary to church teaching was placed on the *Index Expurgatorius* in Rome. His *Hydrotaphia; Urne-buriall*, prompted by the discovery of some sepulchral urns in Norwich, has been described as 'hardly paralleled in the English language for its richness of imagery and majestic pomp of dicta'. A portrait hangs in the Church of St Peter Mancroft.

NOTTINGHAM, Notts.

Industrial city particularly identified with textiles (especially lace), coal and tobacco. About 1833 a local lacemaker, Hooton Deverill, adapted the Jacquard loom to make lace. At Ruddington (4 miles S) the Framework Knitters Museum exhibits a collection of knitting frames. The National Mining Museum at East Retford depicts all aspects of coal mining, including an exhibit showing the evolution of the miners' safety lamp.

The water-pumping station at Papplewick (10 miles N) has two late 19th-century beam engines and is noteworthy for its ornate architecture incorporating tiles and stained glass. The Canal Museum (Canal St), located in an old warehouse, is devoted to the history of the River Trent and related waterways.

Nottingham Industrial Museum (Wollaton Hall) is devoted to the history of Nottingham's industries in general – including bicycles, tobacco, printing, engineering and pharmaceutical products. Close by is Nottingham Natural History Museum, with both local and international exhibits. The history of the lace industry is depicted in the Lace Centre (Severns Buildings).

Birthplace (*c.* 1205) of William of Sherwood, famous medieval logician. He had a peripatetic career, teaching variously at Paris, Oxford, Aylesbury, Lincoln and Attleborough. Roger Bacon, who tended to have a poor opinion of contemporary philosophers, described him as 'one of the famous wise men of Christendom'.

Among modern scientists associated with the city was F.S. Kipping FRS (1863–1959), pioneer of the organic chemistry of silicon, from which modern silicone plastics derive.

NUFFIELD, Oxon.

William Morris (First Viscount Nuffield, 1877–1963) motor-car manufacturer and liberal philanthropist, especially to scientific and medical institutions in OXFORD – is buried at Holy Trinity Church.

NUTLEY, E Ssx

The Nutley Windmill dates from the 17th century: it has been restored and is again operational. Technically, it is an open-trestle post-mill.

O

OAKHAM, Leics.

Although iron horseshoes were apparently used by some German tribes as early as the 2nd century BC, they were not widely used in Europe until the 8th century AD. It was a major development in the use of draught animals, increasing their tractive power, and created a new rural industry. Over the centuries there have been many changes in design and these are reflected in a remarkable collection at Oakham Castle, dating from the 15th century. Its origin is curious: traditionally every peer of the realm passing through Oakham must pay a horseshoe as a forfeit to the Lord of the Manor. In this way more than two hundred shoes have been accumulated.

OBAN, Strath. See TAYNUILT

OCKHAM, Sry

Birthplace of William of Ockham (or Occam) (*c.* 1285–1349), the *Doctor Invincibilis* who, with Duns Scotus (*c.* 1265–1308) and Thomas Aquinas (1225–74), is recognized as one of the three most influential of medieval scholars. Famous for his dictum (Occam's Razor) that 'entities ought not to be multiplied except of necessity'. Roughly speaking, this means that of all possible explanations of a phenomenon the simplest should be chosen. He instigated the metaphysical concept of nominalism, which dominated academic thought in northern Europe in the 14th and 15th centuries. (See OXFORD.) There is a memorial window in All Saints parish church.

OFFALY (HOLYWOOD), Ofly

Birthplace of John Joly (1857–1933), geologist and physicist. (See DUBLIN.)

OFFA'S DYKE

Offa's Dyke is the longest linear earthwork in the British Isles, running from the River Dee to the Bristol Channel. Although the attribution to Offa, King of Mercia (757–96) has been questioned, it is difficult to imagine any contemporary who could command the authority and resources to bring it into being: with justification he described himself as '*rex totius Anglorum patriae*'. The uniform mode of construction throughout its length suggests one master-mind in control. The Welsh, however, were not minded to acknowledge Offa's suzerainty, and the purpose of the wall was to protect English settlers against raids by dispossessed Welshmen.

The dyke consisted of an earth rampart, faced by a ditch, reaching an overall height of about 7 m; it may have been topped with a wooden palisade for additional protection. Like HADRIAN'S WALL, with which Offa would certainly have been familiar, the dyke was not a fortification proper. Rather, it was an extended sentry walk to give temporary protection until troops could be mobilized. Additionally, of course, it clearly marked Offa's frontier.

Overall, the dyke extends for 150 miles but some 30 miles of this is represented by the natural defences of the River Severn around Shrewsbury and the River Wye from a point just north of Hereford. Much of it can be traversed on foot. The Offa's Dyke Path runs from Prestatyn (Clwyd) to Welshpool (Powys). The state of preservation is variable: one of the best sections is around Knighton, where there is an information centre.

OLD BURGHCLERE, Hants.

In Britain, 'Beacon Hills' abound, reminders of a primitive system of signalling by fires in time of emergency. The one 2 miles W of Old Burghclere is exceptional in having at its summit a grave, that of the archaeologist George Herbert, Fifth Earl of Caernarvon (1866–1923), whose family seat was at Highclere Castle. He is famous for his discovery in 1922 of the tomb of Tutankhamun, with its fabulous treasure, in the Valley of the Kings. Within a year of this discovery he was dead, from a mosquito bite that became infected. The superstitious see this as fulfilling the curse laid by the Pharaoh on anyone who violated his grave.

OLDBURY, Glos.

Village on the River Severn: one of the best points to view the tidal phenomenon known as the Severn Bore, a seasonal tidal wave that may reach a height of 2 m. The Oldbury nuclear power station dominates the adjacent low-lying land. It incorporates two Magnox nuclear reactors.

OLDHAM, Lancs.

Cotton mills, driven by steam engines fuelled with local coal, began to appear in the 1770s. The industry thrived after the American Civil War and huge mills built during the next fifty years – with a characteristically ornate style of architecture – still dominate the town. The Saddleworth Museum reflects the history of local industry – based on textiles and textile machinery – and includes a working woollen mill. It is housed in an old mill on the Huddersfield Narrow Canal which at Diggle (6 miles E) has the longest (3 m) and highest (245 ft above sea level) canal tunnel in Britain.

OLDMELDRUM, Gramp.

Birthplace of Sir Patrick Manson (1844–1922). Pioneer of tropical medicine, remembered particularly for his research on insect-borne diseases, especially malaria.

OLD WARDEN, Beds. See BEDFORD

ORFORD, Sflk

The successful outcome of the Battle of Britain in the Second World War resulted in large measure from the existence of a defensive chain of radar installations to detect invading aircraft. The pioneer of radar was Sir Robert Watson-Watt (1892–1973) and his earliest field experiments were carried out in great secrecy, in 1935, on the lonely marshes adjacent to Orford Ness.

ORKNEY

James Copland (1791–1870) was born in Orkney and went to school there before studying medicine at EDINBURGH. He is remembered for his monumental *Dictionary of Practical Medicine*. It took him nearly thirty years to write and finally amounted to 3500 closely printed pages in three volumes.

OTLEY, W Yorks.

In the years of the Railway Mania one of the most dangerous construction tasks was tunnelling. A memorial in the churchyard here to the men who died building the Bramhope Tunnel on the Leeds and Thirsk Railway 1839–45 is a grim reminder of this. It depicts the entrance to the tunnel and the manager's tower. How many men died is not stated but in the building of the Woodhead Tunnel, on the line linking Sheffield and Manchester, 30 men were killed and 150 seriously injured out of a workforce of about 1000.

OTTERY ST MARY, Devon

The parish church has an astronomical clock (early 17th century) resembling that in the Cathedral at EXETER.

OULTON, W Yorks.

A new 2-mile section of the Aire and Calder Navigation was opened here in 1995, the first new canal in Britain for 90 years. It cost £20 million to build, compared with £26,700 for all 48 miles of the original canal, completed in 1704. This carried freight from the mill towns of Leeds, Halifax and Wakefield via the River Humber to the sea.

OXENHOPE, W Yorks.

The Keighley and Worth Valley Railway, now privately owned, possesses vintage locomotives and carriages, including a Pullman carriage. Its museum is adjacent to Oxenhope Station.

OXFORD, Oxon.

Although there were earlier Roman and Saxon settlements, Oxford seems first to have been named as a burgh in 912, when Edward the Elder established control of the Thames Valley by 'taking to London and Oxford and all the lands belonging thereto'. However, this stability was transient, for the town was repeatedly devastated by the Danes in the 11th century. Over the period 1170–1220 the Norman governor Robert d'Oyly created a strong fortress, of which one tower still stands. The completion of this roughly coincided with the influx of scholars from which arose the University, one of the world's great centres of learning.

For centuries Oxford and the University were almost synonymous. Such local industry as there was existed on a small scale, supplying local needs: typical examples were brewing, weaving, pottery and quarrying. Their customers were fellow townsmen and members of the University – numbering fifteen hundred by the early 13th century – and the colleges which began to dominate it. As late as 1899 the antiquary J.M. Falkner, in his *History of Oxfordshire*, confidently asserted that 'Oxfordshire was never fated to be an industrial county, and if we except some tweed-making at Chipping Norton and some implement-making at Banbury, it remains guiltless of manufacture to this day'. The word guiltless is revealing, indicating the traditional attitude of the University to industry. In fact, the seeds of change had already begun to germinate. Already in 1893 sixteen-year-old William Morris (1877–1963), with £4 capital, had set himself up as a bicycle repairer and by 1896 had established himself as a manufacturer of bicycles and motor-bicycles at 48 High Street. Gradually he moved up the industrial scale, becoming a 'motor car engineer' in Morris Garages at 100 Holywell Street. From this it was a small step to manufacturing motor-cars, largely from bought-in components, for the popular market: in 1913 he sold a thousand Morris Oxford cars at £175 each. The war interrupted this promising business but he resumed in 1919 in new premises at Cowley and during the next decade established himself as one of Britain's leading manufacturers, catering for the demand for small family cars. Soon Cowley was outgrown and additional plants were set up in BIRMINGHAM and COVENTRY. He became one of the most successful and wealthiest industrialists of his day, yet preferred to live modestly and devoted much of his wealth to philanthropic ends: it is estimated that his donations totalled more than £30 million, mostly made after the company first issued shares in 1936. In his later years he gained many public honours, notably a peerage in 1938, when he became Viscount Nuffield: it is with this name, rather than Morris, that most of his benefactions are associated. He was elected a Fellow of the Royal Society in 1939. His largesse was widely distributed, but Oxford came in for a substantial share of it. In the early 1930s he financed the building of an Institute for Medical Research adjacent to the Radcliffe Infirmary and the rebuilding of the Wingfield Orthopaedic Hospital. In 1937 he contributed £100,000 towards the building of a new Physical Chemistry Laboratory in South Parks Road. He also gave most timely support to the penicillin project in Oxford in its early stages when its potential was not widely recognized. In 1943 he gave £10 million to establish the Nuffield Foundation which, among other philanthropic donations, has given wide support to medical and scientific research. Among the projects to which it gave assistance from the outset was the great radio telescope built by Manchester University at JODRELL BANK. When the university found itself in serious financial difficulties in 1963 as a result of over-spending on the project, Nuffield personally undertook to meet the deficit. Surprisingly, in view of his munificent support for higher education and research, Nuffield favoured non-graduate management in his own business – with increasingly damaging results – so his great enterprise itself produced few technological achievements.

Despite his generosity, Nuffield's relationship with the University was not entirely happy. In the 1930s he was minded to found a new college devoted to engineering and offered £1 million to endow it. Regrettably, the University as a corporate body was strongly orientated towards the arts and had little enthusiasm for science, let alone the mechanic arts, as a proper subject for academic inquiry. His offer was rejected and instead he was persuaded to endow Nuffield College – his most conspicuous memorial in Oxford – for the study of economic, social and political problems. Due to wartime delays, this was not finally incorporated until 1958. Nuffield, a Conservative by conviction and a life-long opponent of socialism, privately referred to it as the Kremlin, believing its studies to be more socialist than social. Nevertheless, he was obviously reconciled, for he left the College a generous bequest on his death. In retrospect, it is clear that this opportunity to establish an effective interest in engineering was an error of judgement on the part of the University. Although Nuffield was one of Oxford's greatest benefactors there is no public monument to him in the city. His ashes are buried in the churchyard at NUFFIELD, where he made his home in his later years.

Until comparatively recently the various Nuffield enterprises were by far the largest employer in the Oxford area, but – save for the Rover Group – the great Cowley site is now a shadow of its former self, much of it having been razed for redevelopment. Three other industries – two very old and one very new – deserve mention. One is the University Press, in Walton Street. Founded in 1478, it grew to be one of the largest printing works in the country, its prosperity owing much to the granting of a licence to print bibles towards the end of the 17th century. In the centre of Oxford, in the shadow of Robert d'Oyly's castle, is a brewery where the Morrell family has brewed beer for six generations. It is now open to visitors and displays illustrate the history of brewing technology. Exhibits include a working water-wheel and a 19th-century boiling kettle. The new industry is Oxford Instruments, a high-technology company founded in 1959. It manufactures highly sophisticated apparatus for medicine, science and engineering – the latter including equipment for particle detection in high-energy physics. At its peak the Nuffield Organisation employed around twenty thousand people at Cowley and other sites in the Oxford area and dominated the local labour market. For a time it far exceeded the University which for centuries had – directly or indirectly – been dominant. Nevertheless Oxford remained predominantly a university city, as the Cowley works were three miles from the town centre.

No precise date can be assigned to the beginning of the University. By the early 12th century, monasteries had been founded at St Frideswide's and Osney and some informal community of scholars had been established. This was reinforced in 1167 when there was an exodus of scholars from Paris. In 1214 a Papal Charter was granted, making Oxford the recognized university of the north, on a par with Paris and Bologna. Even in these formative years science was regarded as a proper subject for study, for Robert Grosseteste (c. 1168–1253) was appointed first Chancellor. One of the greatest of medieval scholars, with a mastery of human knowledge as it then was, he had a particular interest in astronomy, optics, heat and sound. He was closely followed by Roger Bacon (c. 1214–92), the *Doctor Mirabilis* of his day, often regarded as the first scientist in the modern sense. Like his master, he took the whole of contemporary knowledge as his province. Traditionally, he invented gunpowder, but there is no truth in this: it was an invention imported from China. Perhaps his most important legacy was his teaching that speculation is fruitless in the absence of facts. He was a misanthrope and during his settled years in Oxford (1250–77) worked alone in a house on Folly Bridge, which survived until 1799. He was attached to a friary which is now the site of the Westgate shopping centre: there is a small memorial plaque on the wall of the centre in Old Greyfriars Street.

In its early days the University possessed privileges under its Charter, but no buildings or endowments. Not until the 13th century did the present collegiate system begin, with the foundation of University College in 1249, followed by Balliol (c. 1263) and Merton (1264). The last had from the beginning a small group of scholars with an active interest in science, medicine and mathematics. Its library is one of the oldest in the country and among its prized possessions is an astrolabe that once belonged to

Geoffrey Chaucer (1340–1400). How this was acquired is not known – for Chaucer had no particular Oxford connections – but he certainly had a keen interest in astronomy and alchemy. His *Treatise on the Astrolabe* was published in 1391. Merton had a far-reaching influence on English academic life, for it was adopted as the model for Peterhouse, the first of the great CAMBRIDGE colleges.

Over a period more colleges were founded – four in the 14th century and three in the 15th; six more were founded between 1509 and 1571. This resulted in a complete change of emphasis: real power lay with the self-governing colleges, subject only to the general statutes of the University. Not until the University Acts of 1854 and 1877 did this begin to be reversed. The change was reinforced by the further University Act of 1923, by which for the first time the University received an annual block grant from the Treasury: this now represents the greater part of the University's income. The change was inevitable, because although many of the colleges were well endowed – and a few even had small laboratories – their income was insufficient to support new requirements, especially for teaching and research in science.

Early science in Oxford, as elsewhere in Europe, was largely a perpetuation of the teaching of Aristotle, Pliny, Galen and other scholars of the classical world, much of it passed on – and often reinforced – through Arabic texts translated into Latin. It was essentially dogmatic, and not until the 17th century did science really emerge as a creative activity instead of simply passing on perceived – and theologically acceptable – wisdom. In this new development Oxford was very much in the forefront, for it was here that was gathered the small group of scholars who in 1660 initiated the Royal Society, one of the oldest and the most prestigious scientific society in the world. Some of them had previously been meeting in LONDON about 1645, but their proceedings were interrupted by the Civil War and the Protectorate. In Oxford they found kindred spirits and a congenial intellectual environment.

Merton had maintained its scientific tradition. In 1585 Sir Henry Savile (1549–1622) had been elected Warden and regularly lectured on mathematics. He founded the Savilian Chairs of Geometry and Astronomy and helped his friend Thomas Bodley (1545–1613) to found the famous Bodleian Library. The first Savilian Professor of Geometry was Henry Briggs (1561–1631). His own work was not highly original but he disseminated knowledge of John Napier's invention of logarithms and his trigonometrical tables were used until the early 19th century. Savile has a handsome memorial in the college chapel: Briggs' is modest, a stone slab inscribed simply 'Henricus Briggius'. Briefly – very briefly – Merton had a very distinguished Warden in William Harvey (1578–1657), discoverer of the circulation of the blood. During the Civil War he was in attendance in Oxford on Charles I, who arbitrarily appointed him Warden in 1645: he lost his job less than a year later when Oxford surrendered to Cromwell.

Another college similarly hard-hit was Wadham, thoroughly purged by the Parliamentary Commission in 1648. John Wilkins (1614–72) – brother-in-law to Oliver Cromwell – was appointed Warden and held the office for 11 years before becoming Master of Trinity College, CAMBRIDGE. He survived the Restoration and eventually became Bishop of CHESTER. Wilkins had no great scientific pretensions but he had a great interest in mechanical devices and was a versatile writer: among other things he wrote a book on codes and ciphers. His *Discovery of a World in the Moon* is perhaps the first example of science fiction writing. He did, however, attract round him a coterie of natural philosophers. In the manner of the day, they used to meet at Tillyard's coffee house, on the site of what is now 90 High Street. For a time, Oxford was a hive of scientific activity.

In 1654 Robert Boyle (1627–91) settled in Oxford until moving to London in 1668, and made his home with John Crosse, an apothecary, at his house in the High Street. The site is marked by a plaque on the wall near University College. He is remembered for his work on pneumatics and for enunciating Boyle's Law, relating the volume and pressure of a gas. He is also remembered as 'the Father of Chemistry'. His *Sceptical Chymist* (1662) is one of the great classics of science. In this he moved away from the traditional ideas of the alchemists and developed a new chemistry on atomistic lines. In Oxford he had as his assistant the young Robert Hooke (1635–1703), one of the most brilliant and versatile of 17th-century scientists. He constructed an improved air-pump for Boyle's experiments with gases.

Another member of this scientific group was the mathematician John Wallis (1616–1703), whom Cromwell had appointed Savilian Professor of Geometry in 1649 in spite of his Royalist sympathies. His most famous book was his *Arithmetica Infinitorum* (1655), largely an analysis of surfaces. He derived an elegant arithmetical formula to calculate π and introduced the symbol ∞ to denote infinity.

Also in Oxford at this time was Christopher Wren (1632–1723), initially as a student at Wadham. In 1653 he was elected to a Fellowship at All Souls. Briefly (1657–61) he was Professor of Astronomy at Gresham College, London, before returning to Oxford as Savilian Professor of Astronomy (1661–73). A gifted mathematician, he is generally better known as an architect, most particularly for his rebuilding of St Paul's after the Great Fire of 1666 and of the original Greenwich Observatory. There are many memorials to him in Oxford, including the splendid sundial on the chapel of All Souls. Others are the Sheldonian Theatre and Tom Tower of Christ Church.

Boyle, Wilkins, Wallis, Hooke and Wren were all original members of the Royal Society: Hooke was its first curator and Wren was elected President in 1680. Somewhat later than these was Edmond Halley (1656–1742), who entered Queen's College in 1673 but did not stay to graduate. Instead, he made his reputation by cataloguing the stars of the Southern Hemisphere and his prediction of the periodicity of comets. He identified the comet of 1531 with those of 1607 and 1682 and correctly predicted that it would reappear in 1758, long after his death; it last appeared, undramatically, in 1986. In 1703 he was appointed Savilian Professor of Geometry and succeeded John Flamsteed (1646–1719) as Astronomer Royal in 1720. He also made a notable contribution to science in quite another way, for it was he who persuaded Newton to publish his great *Principia* and paid for its printing. There is a memorial plaque on his house by the Bridge of Sighs in New College Lane, where he also had an observatory.

In parallel with this flourishing of the physical sciences and mathematics, an important botanical event can be recorded. In 1621 the Earl of Danby founded the Oxford Physic Garden; it is the oldest in Europe after Pisa and Leyden. Its purpose was to grow plants useful to medicine and botany and specimens were gathered from all over the world. Among the novelties was *Senecio squalidus*, introduced from Sicily in 1794: in due course it escaped and as Oxford Ragwort is a troublesome weed over much of southern England. The magnificent entrance gateway was built by Nicholas Stone (1586–1647), master-mason to Inigo Jones. In 1834 experimental work was introduced by Charles Daubeny

(1795–1867) when he was appointed Professor of Botany. It is a reflection on the nature of academic life at that time that he already had, and retained, a professorship in chemistry and was subsequently also appointed Professor of Rural Economy. Doubtless this plurality of appointments helped him to build a laboratory at his own expense. Remembering Oxford's long tradition in science he had inscribed over the entrance *Sine experientia nihil sufficienter scire potest* ('Without experience nothing can be sufficiently known'), taken from Roger Bacon's *Opus Majus* written nearly six centuries earlier.

Over the centuries Oxford has had many generous benefactors, and among the greatest was John Radcliffe (1652–1714), a graduate of Lincoln College who became a fashionable physician in London. On his death he left much of his large fortune to the University: he is buried in the University Church of St Mary's in the High Street. Among his major memorials are the Radcliffe Camera in Radcliffe Square, a part of the Bodleian Library, the original Radcliffe Infirmary and the Radcliffe Observatory in Woodstock Road. The last was finally completed in 1799 and was used for astronomical purposes until 1935. It is a striking landmark, for the building, in the classical style, is capped by a version of the Tower of the Winds, the octagonal horologium built in Athens *c.* 100 BC for measuring time. It now forms part of Green College, primarily for medical students and their teachers. It has a strong technological connotation, for its foundation owed much to Dr and Mrs Cecil Green, of Texas Instrument Inc. in the USA.

Generally speaking, the University was at a rather low ebb in the 18th century, and indeed in the early 19th. In 1784, however, both Town and Gown were enlivened by an ascent in a hot-air balloon by the pioneer English aeronaut James Sadler (1753–1828), barely a year after the first flight by the Montgolfier brothers in France in 1783. He made a flight of about six miles; its starting point is marked by a plaque in Deadman's Walk, close by the Botanic Garden. (See also MANCHESTER.)

Oxford is noted for its eccentrics and among them was William Buckland (1784–1856) who entered Corpus Christi in 1801 and was elected Fellow there in 1808. He was ordained, and became a canon of Christ Church in 1825. For twenty years he lived with his family in a corner of Tom Quad, keeping there a considerable menagerie of wild animals and birds, members of which were wont to appear on his dinner table. In 1813 he was appointed Professor of Mineralogy, and later to a readership in geology. He became an enthusiastic collector of fossils and sought to reconcile the record of the rocks with the biblical account of Creation. Of him Philip Shuttleworth, Warden of New College, wrote:

> Some doubts were once expressed about the Flood
> Buckland arose, and all was clear as – mud.

However, clarification did begin shortly after Buckland's death, with the publication of Charles Darwin's *On the Origin of Species* in 1859. Predictably, the theologians were outraged and matters came to a head in the summer of 1860 at a meeting of the British Association in Oxford. There was a famous confrontation between 'Darwin's Bulldog' T.H. Huxley (1825–95) and Samuel Wilberforce (1805–73), Bishop of Oxford; because of his oily and ingratiating manner, he was commonly known as 'Soapy Sam'. Benjamin Jowett (1817–93) of Balliol remarked of him that 'Samuel of Oxford is not unpleasing if you will resign yourself to being semi-humbugged by a semi-humbug'. Deriding the whole concept of organic evolution, Wilberforce offensively asked

Huxley if it was through his grandfather or his grandmother that he claimed descent from a monkey. In a measured and devastating reply, Huxley said that he was not ashamed to be descended from an ape but he would be ashamed of an ancestor who used great gifts and eloquence to obscure the truth.

This confrontation is noteworthy for another reason: it took place in the newly completed University Museum, which had been built to meet the needs of the Honour School of Natural Sciences created in 1850. For a time it housed all the University scientific departments and all later laboratories have been built in the same area. The design was very much influenced by John Ruskin and it is a fine example of Victorian Gothic architecture. The central area, whose glass roof is supported by ornate iron pillars, is surrounded by an ornate arcaded area on two floors. The attached chemical laboratory is bizarre in the extreme, being whimsically modelled on the Abbot's Kitchen at Glastonbury. Nevertheless the Museum did mark a turning-point in the history of science in Oxford. The first keeper of the Museum was John Phillips (1800–74) – a founder of the British Association at York in 1831 – who had succeeded Buckland as Professor of Geology. Immediately adjacent is the Pitt Rivers Museum, one of the world's leading ethnological museums. It was founded in 1883 to house a splendid collection made by General H.L.F. Pitt Rivers (1827–1900). Over the years the collection has been greatly enlarged and the museum now has additional accommodation in the Banbury Road. The collection is remarkably diverse, ranging from fire-making devices to musical instruments, from a large totem-pole to shrunken heads from the Pacific. The museum is also a teaching establishment, part of the Department of Ethnology and Prehistory.

Yet another great benefactor was the 1st Earl of Clarendon (1609–74). As Chancellor he conceived the idea of an academy for instruction in dancing, fencing and riding. This came to nothing, but his great-grandson left all the Clarendon archives to the University, with the thought that they might be sold and the proceeds used to finance such an academy. Sold they were, but the money was left to accumulate and in 1868 was used to build the original Clarendon Laboratory for physics. The first Professor of Experimental Physics was R.B. Clifton (1836–1921) but despite gaining an FRS and listing work as his recreation in *Who's Who*, he achieved little. By his own account 'the wish to do research betrays a certain restlessness of spirit'. In 1919 he was succeeded by a very much more vigorous professor in the person of F.A. Lindemann (1886–1957) better known as Lord Cherwell, scientific adviser to Winston Churchill during the Second World War. Shrewdly offering appointments to gifted refugees from Europe in the 1930s, he built up a strong research team, with a particular interest in very low temperature physics (cryogenics). In 1939 a new laboratory, the Lindemann Building, was completed.

This coincided with a general resurgence of science in Oxford, underlined by a rich crop of Nobel Prizes. In 1919 Frederick Soddy (1877–1956) had been invited to Oxford from GLASGOW to become Dr Lee's Professor of Chemistry. It was hoped that he would found a British school of radiochemistry, and this hope was encouraged when he was awarded a Nobel Prize in 1921 for his discovery of isotopes. However, he became increasingly morose and withdrawn, and preoccupied with political and economic theories which gained no acceptance. His successor, C.N. Hinshelwood (1897–1967) was a man of very different mettle. He established an international reputation for his research on the kinetics of chemical reactions. Initially he worked under very poor conditions in a basement laboratory in Trinity; Queen's, Christ Church and Jesus also maintained small laboratories, the last until as late as 1946, but in 1940 he moved to a splendid new laboratory in South Parks Road. He was awarded a Nobel Prize in 1956, but his talents lay not only in chemistry. He was a gifted linguist, an expert on Chinese ceramics, and was simultaneously President of the Royal Society and of the Classical Association, a double otherwise equalled only by the geologist Archibald Geikie (1835–1924).

Hinshelwood's immediate neighbour in South Parks Road was the organic chemist Robert Robinson (1886–1975), who as Wayneflete Professor had inherited the Dyson Perrins Laboratory in 1930 from W.H. Perkin Jr (1860–1929), son of Sir William Perkin (1838–1907), pioneer of the dyestuffs industry (see MANCHESTER). This was named after the philanthropist and art collector C.W. Dyson Perrins, whose wealth derived from the famous Lea and Perrins Worcester Sauce and who made a very substantial contribution towards the laboratory's cost. Robinson, with a particular interest in the chemistry of natural products, was perhaps the greatest chemist of his day in the classical tradition. Like Hinshelwood, he served as President of the Royal Society, and was awarded a Nobel Prize in 1947. Although relatively new when Robinson arrived, by the time he retired in 1955 it had become dilapidated, mainly as a result of wartime neglect. It was then extended by new buildings and well equipped with the sophisticated apparatus of modern chemistry. The outside of the old building is distinguished by two cryptograms. One simply links the almost identically spelt names of Perrins and Perkin, Robinson's predecessor. The other, much more subtle, cryptogram was devised by the architect, Paul Waterhouse (1861–1924), a Balliol man. It defied Oxford scholars for many years – some came to regard it as a hoax – but was eventually deciphered by D.Ll. Hammick, one of Perkin's colleagues, who, in the old tradition, was versed in Latin as well as chemistry. It proved to be a punning reference to 'Waterhouse of Balliol'.

Another Oxford Laureate of this period was Charles Sherrington (1857–1952), Professor of Physiology 1913–32, famous for his pioneer research on the physiology of the nervous system. He has justly been called the 'Harvey of the nervous system', and was awarded the Nobel Prize in 1932. He, too, served as President of the Royal Society 1920–5.

But the real jewel in the crown of scientific research in Oxford and, indeed, in the world, was the work on penicillin initiated in 1939 in the Sir William Dunn School of Pathology by an Australian, H.W. Florey (Baron Florey of Adelaide and Marston 1898–1968) and E.B. Chain (1906–79), a refugee from Germany. Penicillin had been discovered at St Mary's Hospital in LONDON, by Alexander Fleming (1881–1955) in 1928, but he failed to recognize its unique therapeutic value and had abandoned interest by 1931. Originally, the Oxford group's interest in penicillin was not in its potential medical use but as part of a broad study of the antagonisms between different kinds of micro-organisms. By 1940, however, it was apparent that penicillin had unique potential for the treatment of a wide range of bacterial infections. Potential was the word, however, because the crucial experiment, conducted on 25 May 1940 with four mice, was a very long way from convincing clinical trials and the mass production of penicillin to treat military casualties in the latter part of the war. In the event, this demanded the marshalling of a major industrial crash-programme among pharmaceutical companies in the USA. With radar and the atomic bomb, penicillin was one of the greatest scientific and technological achievements of the Second World War. A stone memorial near the entrance to the Botanic Garden commemorates the ten research workers who contributed to the dramatic success of the penicillin project in Oxford. It, and an associated rose garden, was endowed by the Lasker Foundation of New York, which had given Florey substantial support at a critical moment. In 1945 Fleming, Florey and Chain shared a Nobel Prize. Florey went on to a life peerage and the Presidency of the Royal Society 1960–65. A memorial in Westminster Abbey, near those of the astronomers John and William Herschel and the naturalist Charles Darwin, reads simply: 'His vision, leadership and research made penicillin available to mankind.'

These successes continued after the war. In 1964 Dorothy Hodgkin (1910–94) became only the third woman to be awarded a Nobel Prize – the others were Marie Curie (1903) and Irène Joliot-Curie (1935) – for pioneer work on X-ray crystallography, including evaluation of the structure of penicillin. She was a graduate of Somerville, the non-denominational college for women founded in 1879: it was named after Mary Somerville (1780–1872), a mathematician of distinction and a member of the benefactors' family.

Another Nobel Laureate in the 1950s was Hans Krebs (1900–81), famous for his elucidation of the metabolic pathway known as the Krebs cycle. However, he had won the prize in 1953, the year before he came from SHEFFIELD to Oxford as Professor of Biochemistry. His successor, Rodney Porter (1917–85), was awarded a prize in 1972 while still in office. In the following year the Dutch-born Nikolaas Tinbergen (1907–88), founder of the new discipline of ethology (animal behaviour), was similarly honoured. He had been appointed Professor of Animal Behaviour in 1966. It was a family triumph: his brother Jan had already won a Nobel Prize for economics in 1969.

All in all, it was half a century of high scientific achievement in Oxford: nine Nobel Laureates, the highest world honour, three of whom were also Presidents of the Royal Society, the highest national honour for a scientist.

Despite much commercial development, the centre of Oxford is still very evidently an ancient university city. A major reason for this is that most of the early foundations were located for security within the confines of the old city walls, leaving rather little space for what is now called in-filling. Amidst this conglomerate are two famous museums. The greatest is the Ashmolean, named after Elias Ashmole (1617–92), whom Anthony Wood described as 'the greatest virtuoso and curioso that ever was known of or read of in England before his time'. It was based on a remarkable collection acquired – by rather dubious means – from the Tradescants, famous gardeners of Lambeth in LONDON. It was originally housed in the Old Ashmolean Building in Broad Street but moved in 1845 to the University Galleries in Beaumont Street, where it was reinforced by other University treasures from the Bodleian Library and elsewhere. It is now a marvellous collection of international reputation, but of no particular scientific interest. The Old Ashmolean served for a time as a chemical laboratory, but since 1924 has been the Museum of the History of Science. Largely, but not entirely, it displays instruments, especially ones with astronomical and mathematical connotations. It also displays some pieces of equipment from the Oxford penicillin project, and the X-ray spectrometers used by H.G.J. Moseley after he moved from CAMBRIDGE to Oxford in 1913. It also displays the blackboard used by Einstein in 1931 during an exposition of the coefficient of the expansion of the Universe in terms of its mean density. The Museum of Oxford, in St Aldates, records the history of Oxford from its earliest days to its transformation into an industrial city.

P

PADDOCK WOOD, Kent

The Whitbread Hop Museum displays farm machinery and various rural crafts (wheelwright, blacksmith, dairy, etc.).

PAIGNTON, Devon

Oliver Heaviside (1850–1925) is buried in Paignton Cemetery. He is famous for identifying the Heaviside layer in the upper atmosphere, which reflects radio waves and prevents their escaping into outer space.

PAISLEY, Strath.

Once the main centre of the cotton industry in Scotland, world famous for its shawls, a fine collection of which is exhibited in the Paisley Museum and Art Galleries. At Kilbarchan (4 miles W), a typical weaver's cottage (NT) has been restored and the loom of a local weaver and all its accessories has been installed in the basement.

Birthplace of Robert Broom (1866–1951), palaeontologist. Also of William McNaught (1813–81), inventor of the compound steam engine.

PAPPLEWICK, Notts. See NOTTINGHAM

PARYS MOUNTAIN See AMLWCH

PAYHEMBURY, Devon

A splendid hillfort begun around 4000 BC and enlarged and improved until the start of the Christian era. Located 1 mile S of A373 Cullompton–Honiton Road.

PEEL, IoM

The round-tower is one of only three known outside Ireland (cf ABERNETHY).

PENDEEN, Corn.

The once flourishing Cornish tin-mining industry is now nearly extinct. Geevor Tin Mine is the last one still working and it has a museum depicting the history of the industry.

PENKULL, Staffs.

Birthplace of Sir Oliver Lodge (1851–1940), chiefly remembered for his research on the propagation of electromagnetic waves (see OXFORD).

PENRITH, Cumb.

Birthplace of the ironmaster John Wilkinson (1728–1808) (see BERSHAM). Penrith Steam Museum is housed in buildings occupied by three generations of an agricultural engineering firm. On view are many items of equipment, including steam engines, foundry and pattern shop.

Long Meg and her Daughters (5 miles NE) is one of the largest stone circles in Britain: two of the stones each weighing nearly 30 tons. It much impressed Wordsworth:

> A weight of awe, not easy to be borne,
> Fell suddenly upon my spirit . . .
> When first I saw that family forlorn.

PENSHURST, Kent

Home from 1856 until his death of the manufacturing engineer James Nasmyth (1808–90). At his works in MANCHESTER he built over three hundred steam locomotives and devised many original machine tools. His most notable invention was the steam hammer, originally developed to forge the giant driving-shafts for I.K. Brunel's *Great Britain* (see BRISTOL). In 1856 he retired to Penshurst, a wealthy man, to enjoy his hobby of astronomy, especially mapping the moon's surface. His steam hammer and his lunar map were both exhibited at the Great Exhibition in 1851. The house he bought he aptly renamed Hammerfield, not only after his own invention but from 'hereditary regard for hammers, two broken hammer-shafts having been the crest of the family for hundreds of years'.

PENZANCE, Corn.

Humphry Davy, the chemist who later achieved international fame at the Royal Institution in LONDON, was born here in 1778. For a time he was apprenticed to a local surgeon and carried out his first chemical experiments. In 1798 he moved to BRISTOL as assistant in the Pneumatic Institution. A commemorative statue stands at the top of Market Jew Street.

The area is rich in remains of archaeological interest. At Chysauster are the ruins of a courtyard village dating from about 100 BC and occupied during Roman times. Tin was mined there.

Lanyon Quoit at Madron is a Neolithic tomb from which the covering soil has disappeared. Three massive uprights support a huge capstone 6 m long. The Mem-an-Tol (Stone with a hole in it) is the central block of three: legend says that sick children were treated by being passed through the hole. The Merry Maidens at St Buryan (5 miles SW), one of the most perfect stone circles in Britain, is 25 m in diameter. It comprises nineteen large rectangular blocks and dates from the Bronze Age. At Treen (7 miles SW) is a hillfort with five ramparts and a famous megalith, the rocking Logan Stone.

PERTH, Tays.

Home of Adam Anderson (d.1846) physicist, one time Rector of Perth Academy and later of St Andrew's University. His research was not noteworthy but the building he designed in 1832 for the Perth Waterworks (now an information centre) is a memorial to him; designed in the classical tradition, with a domed roof, it is a fine early example of building construction in cast iron.

At Stanley (6 miles N) is a monument to the social reformer Richard Arkwright (1732–92); it is a complex of large stone-built mills, started in 1785, and an associated village for the workers. Arkwright is remembered also for a similar enterprise at NEW LANARK.

Pullars of Perth, well-known dyers, in 1856 confirmed that W.H. Perkin's mauveine – the first synthetic dye – had commercial potential, thus founding an important new branch of the chemical industry.

PETERBOROUGH, Cambs.

The Nene Valley Railway exhibits a large collection of industrial and main-line locomotives at Wansford Station. Some are used to operate a passenger service on a 5-mile track. The Country Centre at Sacrewell Farm (8 miles W) exhibits a working 18th-century water-mill on a Domesday site, together with a large collection of agricultural machinery and equipment.

PETT'S WOOD, Kent

Here the National Trust owns some 130 acres of woodland. A granite sundial at the north corner (Orpington Road, Chislehurst) commemorates William Willett (1856–1915), a local resident and leading campaigner for Daylight Saving.

PETWORTH, W Ssx

Petworth House (NT) has a collection of scientific instruments accumulated by Elizabeth Ilive, mistress of the 3rd Earl of Egremont. It includes a large terrestrial sphere, dated 1592, by Emery Molyneux (fl. 1587–1605), 'the rare artizan'. Reputedly it was given to Henry Percy, 9th Earl of Northumberland, by Sir Walter Ralegh when both were prisoners in the Tower of London.

PILTDOWN, E Ssx

Site of one of the most famous of all scientific hoaxes – the skull of Piltdown Man, discovered 1912/13. It was eventually proved to comprise a modern human cranium and the jawbone of an orang-utan. Appropriately, there are two memorials: one lauding the discovery, the other explaining the hoax.

PITCHFORD, Shrops.

The name of this village is significant. Here, in 1697, 'British Oil' began to be produced by Martin Eele by extracting a local bituminous shale with boiling water. The resulting pitch was sold to Severnside boatbuilders for caulking. It had the advantage over wood pitch of not becoming brittle with age.

PITLOCHRY, Tays.

Schiehallion (1083 m) is famous as the mountain chosen – because of its unusually symmetrical shape – by the Astronomer Royal Nevil Maskelyne (1732–1811) to measure the force of gravity by the deflection of a plumb line (1774).

PITMEDDEN, Gramp.

Pitmedden is famous for its Great Garden, laid out in the formal 17th-century French style. The Museum of Farming Life is based on a collection of domestic and agricultural equipment donated by a local farmer and since augmented.

PITSTONE, Bucks.

The outstanding fine post-mill at Pitstone Green dates from 1627. Pitstone Green Farm Museum is located on a working farm: it displays a collection of agricultural and domestic equipment, mostly local.

PITTENWEEM, Fife

The Lochty Private Railway operates steam-hauled trains on its own track. The locomotives comprise engines by several manufacturers and the freight wagons were used in Fife 1900–75.

PLYMOUTH, Devon

Seaport, long an important naval base. Devonport Dockyard and the Dockyard Museum are open to the public.

The third Eddystone Lighthouse was completed by John Smeaton (1724–92) in 1759 and stood 21 m above the rock. In 1880 the sea water was found to have undermined it; the top half was dismantled and re-erected on Plymouth Hoe.

The Royal Albert Bridge over the River Tamar at Saltash, opened in 1859, was built by Isambard Kingdom Brunel (1806–59) to carry the Great Western Railway into Cornwall.

POLEGATE, E Ssx

Polegate tower mill was built in 1817 and continued in use until the Second World War. It was restored to working order in 1967. The adjacent storeroom has been converted to a museum of milling.

POLZEATH, Corn.

The Rumps (NT) on Pentire Head is a cliff fort protected by three concentric ramparts. It was occupied 100 BC–AD 100 and pottery remains indicate regular trade with France.

PONTERWYD, Dyfed

The Llanwernog Silver-Lead Mine Museum (on A44 Aberystwyth–Llanidloes road) – in the heart of an old lead-mining area – is associated with a Victorian mine powered by water. Much of the old mining machinery and equipment – including a water-wheel – has been well restored and is displayed in the open air. A short length of an underground gallery has been made good.

PONTYPOOL, Gwent

Around 1700, 'Welsh Lacquer' goods – mainly household – began to be produced here by coating metal plate with bitumen, derived from local pitch wells (see PITCHFORD), and then stoving them at a high temperature. Later, the process spread to the Midlands where papier mâché was used in place of sheet metal. (See WOLVERHAMPTON.)

Park House was the home of the Hanbury family, who from the late 17th century did much to develop Torfaen as an industrial centre: it is now the responsibility of the Torfaen Museum Trust. It displays the history of local industries, notably iron, steel, tinplate and coal. Junction Cottage (Pontymoel) – a tollkeeper's cottage – is located at the junction of the Monmouthshire and Brecon and Abergavenny Canals. Also managed by the Torfaen Museum Trust, it contains an exhibition illustrating the history of the local canal system.

PONTYPRIDD, M Glam.

For its time, the single-arch bridge over the River Taff, built in 1756 by William Edwards (1719–89) is remarkably graceful. Now bypassed, it is scheduled as an ancient monument. The adjacent chapel – itself an impressive building – is now a museum depicting life in the town when it was a great iron-working centre in the 19th century.

PORT ERIN, IoM

The Isle of Man Railway Museum is associated with a small working steam railway. The unique locomotive *Caledonia* was built in 1885.

PORTHMADOG, Gynd. See TREMADOC

PORTSMOUTH, Hants.

One of Britain's greatest naval bases, with many surviving examples of its past. The Mary Rose Exhibition (HM Naval Base) shows the *Mary Rose* which keeled over and sank in 1545 and was raised in 1982. Also on view are Nelson's *Victory* and HMS *Warrior* (1860), the world's first iron-hulled battleship. The Royal Naval Museum is devoted to all aspects of the history of the Royal Navy.

The Eastney Industrial Museum – in an old pumping station – houses two Watt beam engines and two Crossley gas engines.

Birthplace of Isambard Kingdom Brunel (1806–59), engineer.

PORT SUNLIGHT, Mers.

When W.H. Lever (1851–1925), later Lord Leverhulme, moved his Warrington soapworks in 1888 to a new site on Merseyside he built around it a new township for his workers, thus following in the footsteps of other enlightened industrialists such as Richard Arkwright, Robert Owen and the Cadbury brothers (see NEW LANARK and BIRMINGHAM).

PORT TALBOT, W Glam.

Until comparatively recently, coal was the main industry of South Wales. Its history is illustrated in the Welsh Miners' Museum, Afan Argoed Country Park, which includes a replica of a mine.

PRESCOT, Mers.

The Prescot Museum of Clock and Watch-making displays every aspect of the history of this craft, closely identified with the region from the early 17th century until quite recently (see LIVERPOOL). Exhibits depict successively the traditional hand-crafting of timepieces; a reconstruction of part of the Lancashire Watch Company's steam-powered factory (1899–1910), and modern manufacturing methods, including electronic clocks and watches.

PRESTON, Lancs.

Birthplace of Sir Richard Arkwright (1732–92), pioneer of mechanical spinning. (See NEW LANARK.)

PRESTONGRANGE, Loth.

The Scottish Mining Museum displays various relics of the mining industry but by far its most striking feature is the Cornish beam pumping engine constructed by Harvey's of Hale (1874).

PWLLHELI, Gynd

At Llanaelhaearn (10 miles N) is Tre'r Ceiri, one of the most impressive of all British hillforts. The walls are largely intact and enclose some 150 stone huts, probably occupied AD 150–400 but possibly earlier.

Q

QUAINTON, Bucks.

Quainton is a station on the old Great Central Metropolitan line. The Buckinghamshire Railway Centre was established by local enthusiasts in 1969 to preserve pre-1900 locomotives, rolling stock and other items. A number of locomotives have been restored to operational use.

R

RAMSGATE, Kent

The Ramsgate Motor Museum, originally a theatre, has a considerable collection of vintage cars, motor-cycles, bicycles and memorabilia.

The Spitfire and Hurricane Pavilion (RAF Manston) displays well-conserved aircraft and memorabilia of the Second World War.

RASHARKIN, Ant.

Craig's Dolmen (3 miles N) is a dolmen consisting of a very large capstone and seven uprights.

RATH OF FEERMORE, Gal.

The elaborately carved Turoe Stone, shaped like a giant puffball a metre high, originally stood here but has been moved a short distance away.

RAVENGLASS, Cumb.

The Muncaster Mill occupies a site used from the 15th century. It is unusual in that its overshot wheel is driven by water brought overland from the River Mite, nearly a mile away.

The Railway Museum is housed in the former station of the Furness Railway. The narrow-gauge Ravenglass and Eskdale Railway was built in 1876 to serve the local iron-ore quarries and mines. Today, steam-hauled trains convey passengers from the coast to Eskdale along a 7-mile track. The museum displays many railway memorabilia.

RAYNHAM, Nflk

In 1687 Charles Townshend, 2nd Viscount Townshend (1674–1738) succeeded to his family estate at Raynham. He had enjoyed a distinguished political career but in 1730 retired to devote himself to agricultural improvement. He earned the nickname 'Turnip' Townshend because he introduced the turnip – then essentially a garden crop – into a four-year system of crop rotation. This had revolutionary consequences: it made it much easier to overwinter livestock and avoided the necessity to keep one-third of the land fallow each year. Robert Walpole, whose sister he married, powerfully advocated his methods.

READING, Berks.

Reading's main industries were traditionally the three Bs – beer, biscuits and bulbs. The Blake's Lock Museum (Gasworks Rd) concerns itself with a range of lesser industries – brickmaking, printing, shoemaking and iron-founding among others.

The long association with the manufacture of biscuits by Huntley and Palmer is illustrated in the Reading Museum and Art Gallery (Blagrave St).

The Museum of Rural Life (The University, Whiteknights) is a collection of national status. It depicts the history of farm vehicles, field machinery and other equipment over the last two hundred years.

REDDITCH, H & W

The town has long been famous for the manufacture of needles and fish-hooks. The Needle Museum, Needle Mill Lane, shows fully-restored needle-making equipment dating from 1730. The history of fishing-tackle manufacture is also displayed.

REEDHAM, Nflk

The imposing Berney Arms Windmill, dating from 1840, is a seven-storey mill built for drainage in the Halvergate Marshes.

REETH, N Yorks.

Swaledale Folk Museum has a section dealing with the locally important lead industry.

RENDCOMB, Glos.

Birthplace (1918) of Frederick Sanger, famous for his elucidation (1955) of the chemical structure of insulin. He was awarded a Nobel Prize in 1958 and again in 1980, the latter for his research on DNA. This made him only the fourth person ever to have been awarded two Prizes.

RETFORD, Notts.

Among the exhibits at the National Mining Museum is a 'starvationer', one of the long narrow boats used for conveying coal and other goods on the early canals. Supposedly They were so called because the ribs of the hull were exposed. The other major exhibits include winding and pumping engines, miners' lamps and safety equipment, coal-cutting machinery, and industrial locomotives.

RHOSSILI, Gower

Here, in 1823, William Buckland (1789–1856), the geologist, discovered the famous 'Red Lady of Paviland', a Palaeolithic skeleton. It ultimately proved to be that of a young man buried beneath a mound of red ochre.

ROBERTSTOWN, Kild.

The Canal and Transport Museum is devoted to the history of canals in Ireland, especially the Grand Canal linking Dublin with the River Shannon, the longest river in the British Isles. Work began in the 1770s and ended in 1803. It was unusual in that a series of hotels were built alongside the canal for the benefit of passengers. The Museum is housed in the old hotel at Robertstown.

ROCHDALE, GM

Manufacturing town 12 miles north of Manchester, its prosperity deriving from the textile industry. Ellenroad Mill displays the largest and most complete steam mill engine surviving.

ROCHESTER, Kent

The Kenneth Bills Motor Cycle Museum displays over a hundred machines.

ROLVENDEN, Kent

H.F.S. Morgan built the first of his three-wheel runabouts in 1909. Ten such cars (1913–35) are displayed here at the C.M. Booth Collection of Historic Vehicles (High St), together with the only surviving Humber Tri-car of 1904. A range of other cars, bicycles and tricycles are also shown, all in working order.

ROMAN ROADS

The road system was central to the administration of the Roman Empire. It served not only for the movement of troops and goods, but for the rapid transit of mail. The *cursus publicus*, restricted to the use of government officials, provided relays of wheeled vehicles furnished by post-houses at intervals of about twelve miles. There were also rest-houses to provide overnight accommodation. Private citizens could use the roads but had to make their own arrangements, using inns. The roads were systematically marked with milestones for the convenience of travellers and also to define the sections required to be maintained by the local populace. They were also the basis of itineraries for travellers prepared in Rome.

When they arrived in AD 43 the Romans discovered a land not without well-defined land routes, indicating regular trade. There were, for example, the Ridgeway, running from Beacon Hill (Bucks) to Overton Hill (Wilts), and the Icknield Way, linking Salisbury Plain with the East Coast. Some, like the Icknield Way, the Romans adopted. Other roads they built from scratch to satisfy their own needs. Initially, they radiated from COLCHESTER but soon London was the centre. Typically, they ran straight across country, deviating only when natural features made this impracticable. Among the best-known Roman roads are Watling Street (now the A5), linking London with Chester; Ermine Street and its continuation Dere Street, leading up to the ANTONINE WALL; and the Fosse Way, part of the road from Exeter to Lincoln. Soon there was much in-filling with a network of secondary roads.

The typical Roman road was laid on compacted earth, and consisted of a lower layer of small stones with larger slabs above. While the lines of most Roman roads are well-known, having been adopted for modern use, exposures are rare (but see, for example, TARRANT HINTON and EAST GRINSTEAD).

ROSS-ON-WYE, H & W

Toggles, connected with string loops, were used for fastening garments as early as 2000 BC. The world's leading button museum – displaying over 10,000 items – is in Ross-on-Wye (Kyrle Street).

ROTHAMSTED, Herts.

Rothamsted Experimental Station is an internationally famous agricultural research station. It was founded by Sir John Bennet Lawes FRS (1814–99) after inheriting his family estate there in 1834. His close colleague from 1843 was J.H. Gilbert (1817–1901) who in the last six years of his life was Professor of Rural Economy at OXFORD. In the 19th century it played a leading role in elucidating the relationship between soil composition and the growth of crops.

ROTHBURY, Northld

Cragside (NT) was built by William Armstrong (1810–1900) first Baron Armstrong. A Newcastle engineer, he made a fortune from armaments and warships. Later, his interest turned to electricity and Cragside was the first house in the world to be illuminated by hydro-electric power. The house contains many memorabilia.

ROTHESAY, Strath.

Birthplace of Sir William MacEwen (1848–1924), a student of Joseph Lister and his dresser at GLASGOW. With him he was a pioneer of the technique of aseptic surgery.

The Bute Museum includes material excavated from the Glenvodean burial cairn and from the Iron Age fort at Dunagoil. One section is devoted to the history of River Clyde steamboats, illustrated by some forty models. These include one of the *Comet*, built by Henry Bell (1767–1830) in Glasgow in 1812 for a ferry service between Glasgow, Greenock and Helensburgh. Two years later he had five similar vessels running services on the Thames as far down as Margate.

RUDDINGTON, Notts.

The Framework Knitters' Museum displays, in a setting of old cottages, a collection of frameworks and early textile machinery.

RUDSTON, Humb.

In the churchyard here is set the tallest standing stone in Britain, dating from about 2000 BC. It is 8 m high and 1 m thick and weighs about 26 tons; it appears to have been quarried on the coast at Carnelian Bay, 10 miles away.

RUGBY, Warks.

Birthplace of Sir Joseph Lockyer (1836–1920), pioneer of astrophysics. In the solar eclipse of 1868 he obtained spectrographic evidence of the existence of a new element in the Sun and named it helium. In 1895 Sir William Ramsay (1852–1916) isolated it from the earth's atmosphere.

RUTHERFIELD GREYS, Oxon.

A donkey wheel for raising water from a well survives in the Tudor wheelhouse of Greys Court (NT). (See also CARISBROOKE.)

RYDE, IoW

Cothey Bottom Heritage Centre displays the Norman Ball Transport Collection of road vehicles (1902–70). It includes a number of steam-driven vehicles.

RYE, E Ssx

The fine clock in St Mary's Church dates from the 14th century. It was installed in 1560 but possibly came from Battle Abbey at the time of the Dissolution. The maker was probably the same as that for the Cathedral clocks in EXETER and WELLS.

RYHOPE, T & W

The Pumping Station of the Sutherland and South Shields Water Company was operational from 1869 to 1967 and has since been maintained in working order. The main exhibits are two beam engines built locally in 1869 and still occasionally put into steam.

S

ST ALBANS, Herts.

Francis Bacon (Baron Verulam, Viscount St Albans 1561–1626) is buried at St Michael's Church. His memorial is a life-size figure, seated and seemingly asleep. As a statesman he rose to high office under James I but fell from power in 1621 after conviction for corrupt practices. He believed that humanity would benefit enormously if all human knowledge was systematically categorized and applied. To this end he embarked on a 'grand instauration of the sciences', never fully completed. He argued against the traditional separation of rational and practical knowledge. His so-called Baconian method involved inductive reasoning, a progression from less to more general propositions.

Kingsbury Water Mill on the River Ver dates from the 17th century. Exhibits include the water-wheel, and old agricultural machinery and dairy equipment.

Mechanical musical instruments are a minor, but fascinating, branch of technology. A large collection, based on one accumulated by Charles Hart (d.1983), is displayed at the Organ Museum (Camp Rd), including a ten-bank Wurlitzer.

The Verulamium Museum (St Michael's) is associated with the site of the old Roman city of Verulamium, and the Roman and Iron Age artefacts displayed are among the finest outside London. They include mosaics and wall paintings.

ST AUSTELL, Corn.

This town has been a centre of the china clay industry since 1745 when William Cookworthy (1705–80) of Plymouth discovered the particular merits of certain local clays for the manufacture of porcelain. Today it is used in papermaking, the formulation of medicaments and for making plastics. The vast heaps of waste dominate the local landscape. The Wheal Martyn Museum at Carthew (5 miles N) demonstrates the whole process of mining and refining the clay: exhibits include a 10 m water-wheel.

The docks at Charlestown were built by John Smeaton (1724–92) in the 1790s to accommodate shipping for the export trade.

William Gregor (1761–1817), amateur chemist and mineralogist, was Rector of CREED (10 miles SW) from 1793 until his death.

Antony Hewish, pioneer of radio astronomy and Nobel Laureate 1974, was born (1924) at FOWEY (5 miles E).

ST CLEER, Corn.

Trethevy Quoit is a Neolithic burial chamber from which the covering mound of soil has disappeared. Seven uprights, some 5 m tall, support a huge capstone.

ST HELENS, Mers.

Because of the local availability of the essential raw materials – coal, salt, limestone – Merseyside early became a major centre for the alkali industry. This in turn attracted other industries requiring soda, notably glassmakers. In 1826, William Pilkington (1800–72) became a partner in the newly formed St Helens Crown Glass Company, initially to provide glass for the booming building industry, especially after the excise duty was abolished in 1845 and window tax in 1851. By processes of accretion and expansion this developed into the now-dominant Pilkington Brothers, particularly noted for its development of the float glass process, in which sheet is made by floating molten glass on liquid tin. The firm has a Museum of the Glass Industry in Prescot Road.

St Helens Museum and Art Gallery (College Road) has exhibits depicting mining life.

Birthplace of John William Draper (1811–82), pioneer of the use of photography in astronomy; in 1839/40 he took the first photographs of the moon. By training, he was a chemist, studying at London University; he emigrated to the USA in 1832.

ST JUST, Corn.

At the Geevor Mine, Trewellard (on the St Just–Zennor road) the oldest surviving Cornish steam engine (operational 1840–1930) can be viewed.

ST LYTHANS, S Glam.

The Maes-y-Felin (also known as the Greyhound Kennel) is a large burial chamber from which the soil has disappeared. Three massive uprights support an enormous capstone.

ST MARGARET'S HOPE, Ork.

The Orkney Wireless Museum reflects the work of the communications centre of the naval base at Scapa Flow. The exhibits illustrate the development of radio equipment from crystal receivers to the transistor.

ST MARY'S, IsS

A number of megalithic tombs are located on the island. They include Bant's Cairn, two tombs at Innisidgen and a group of Neolithic chambered tombs on Porth Hellick Down.

ST PETER, Ch. Is.

The Jersey Motor Museum possesses a wide range of veteran and vintage cars, including Allied and German military vehicles of the Second World War.

SALFORD, GM

The Salford Mining Museum depicts the history and technology of coal mining. Exhibits include a collection of miners' safety lamps.

SALISBURY, Wilts.

Mechanical (weight-driven) clocks began to appear in the 13th century. The oldest surviving example in Britain of a clock striking the hours is to be found in Salisbury Cathedral (1386), though it has been substantially modified over the years. (See also DOVER.)

SALTASH, Corn.

The local Cotehele Museum (St Dominick) is an outpost of the National Maritime Museum, Greenwich, and illustrates the history of the local shipping and shipbuilding industries. The chief exhibit is the *Shamrock* (1899) which sailed until 1970 and then became semi-derelict. She has been restored by the National Trust: the Quay itself has been restored to its 19th-century condition.

SANDOWN, IoW

The Museum of Isle of Wight Geology contains some twenty thousand specimens of rocks and fossils. A timely display shows how the island has, over geological time, suffered an alternation of hot and dry and cold wet climates.

SANDTOFT, S Yorks.

The Sandtoft Transport Centre, located on a former wartime airfield, is particularly concerned with the preservation and operation of trolley-buses, now largely out of favour because of the difficulty of fitting them into the modern pattern of urban traffic. Poles and overhead fittings have been obtained from a number of municipal undertakings. The exhibits, now comprising some sixty vehicles from Britain and the Continent, also include motor buses.

SCARBOROUGH, N Yorks.

Sir George Cayley (1773–1857) lived from 1792 at Brompton Hall (on A170 Scarborough–Pickering). He was a pioneer of aeronautics and their indebtedness to his theories of flight was acknowledged by the Wright brothers. Birthplace of William Crawford (1816–95), a founder of palaeobotany. He practised medicine in MANCHESTER and was also Professor of Botany there in Owens College 1880–92. He did outstanding original work on the fossil plants of the Coal Measures which shed much light on the anatomy of many modern plant families.

The 300-ft cliff railway, built in 1875 and working on the water-balance principle, was the first such seaside railway to be built in Britain.

At Pike of Stickle in Langdale there is archaeological evidence of the existence, from as early as 3000 BC, of a remarkable factory for the manufacture of stone axes from the local tuff, which gives a keen edge. At its height as many as three thousand workers may have been involved, exporting axes to all parts of Britain.

SCOUSBURGH, Shet.

At Sumburgh Head is the ancient settlement of Jarlshof, whose seven different occupation periods – ranging from the Neolithic to the medieval – have been identified by excavation. Perhaps the most interesting feature is the ruins of a Viking farmstead.

SCREEN, Wex.

Garrylough Mill (on L29 Gorey road) is a well-preserved corn mill, housing a museum of milling.

SEASCALE, Cumb.

Grey Croft is a large circle of ten stones (reconstructed 1949) close by Calder Hall nuclear power station. Probably early Bronze Age.

SELKIRK, Bdrs

Halliwell's House Museum is unusual in being concerned primarily with ironmongery, of which it has an outstanding collection. It is housed in a reconstruction of an ironmonger's shop.

SENNYBRIDGE, Powys

Cerrig Duon is a circle of twenty-two stones set out much like that at AVEBURY, with far smaller stones. Just outside, however, stands the massive Maen Mawr, fully 2 m tall.

SHEFFIELD, S Yorks.

City world-famous for the making and working of steel – especially cutlery – and the area is rich in reminders of this. The Abbeydale Industrial Hamlet recreates an 18th-century scythe works, featuring a Huntsman crucible works, tilt-hammers and forges. The Kelham Island Industrial Museum features a 12,000 HP steam-engine and illustrates many aspects of Sheffield's industrial heritage. The Shepherd Wheel Museum displays water-powered grinding of cutlery. Wortley Top Forge (10 miles N) is an old ironworks with substantial remains of buildings and machinery. At ELSECAR (9 miles N), a Newcomen engine of 1795 still stands in its original building.

Four technologists are particularly identified with Sheffield. Benjamin Huntsman (1704–1776) settled in Handsworth in 1794 and there developed his famous crucible process for the manufacture of high-quality steel. In 1770 he moved to new premises in Attercliffe and is buried in the churchyard there. The Royal Society elected him to its Fellowship but he declined the honour.

Thomas Bolso(u)ver invented Sheffield Plate in 1742: a thin layer of silver is fused to a thicker sheet of copper and this is then rolled to make a much thinner sheet. By 1774 there were sixteen plating firms in Sheffield alone and the industry had spread to BIRMINGHAM. By about 1850, however, the industry declined with the invention of electroplating.

Harry Brearley (1871–1948) was a self-taught metallurgist who joined Thomas Firth and Company in 1883 as a bottle-washer in the laboratory. He is remembered for the commercial development of stainless steel, especially for cutlery. He left a curiously titled autobiography *Knotted String*, so-called because throughout his life he assiduously collected odd lengths of string and wound them up into huge balls.

Sheffield has been closely identified with development of alloy steels. Henry Clifton Sorby FRS (1826–1908), metallurgist and microscopist, lived his whole life here. Of independent means, he was born into a steel-making family: one of his ancestors was Sheffield's first Master Cutler. He broke new geological ground in using the microscope to examine very thin sections of rocks and later applied the same techniques to investigate the fine crystalline structure of ferrous metals. Sir Robert Hadfield FRS (1858–1940) was closely identified with the development of tough manganese steels. He was the first chairman of the famous firm of Hadfield Ltd, with works at Tinsley.

The city was the birthplace of the civil engineer Sir John Fowler (1817–98), remembered for his work on the London Underground railways and the construction (1882–90) of the great Firth of Forth railway bridge. James Chesterman, inventor of the spring tape, is remembered by a plaque at Bow Works, Pomona Street. Another distinguished citizen, in a very different category, was Samuel Plimsoll (1824–98) – 'the sailors' friend' – who conceived the idea of the Plimsoll mark (see BRISTOL, LONDON). He worked here for a time as clerk in a brewery: there is a memorial plaque at the Sir Harry Stephenson Building, Sheffield University.

Among scientists closely identified with Sheffield was William E.S. Turner (1881–1963) who came to the University as a lecturer (later professor) in 1904. He spent his working life there and in 1925 founded a highly specialized Department of Glass Technology which quickly became the world leader in this field. Another was H.W. Florey (Baron Florey of Adelaide and Marston 1898–1968), who was Professor of Pathology here 1931–5 before moving on to OXFORD where he achieved fame, and a Nobel Prize, for his work on the development of penicillin. A similar career path was followed by Sir Hans Krebs (1900–81), biochemist famous for his elucidation of a basic metabolic cycle (the Krebs' cycle) for which he was awarded a Nobel Prize in 1953. He was at Sheffield 1953–54 before going to OXFORD as Professor of Biochemistry.

SHERBORNE, Glos.

Birthplace of James Bradley (1693–1762), appointed Astronomer Royal in 1742. Famous for his discovery of the aberration of light.

SHIFNAL, Shrops.

Birthplace of the chemist Thomas Beddoes (1760–1808) famous for the foundation in 1792 of his Medical Pneumatic Institute in BRISTOL. There one of his first assistants was the young Humphry Davy (1778–1829) who went on to make a great name for himself at the Royal Institution in LONDON.

SHILDON, Dur.

The Timothy Hackworth Railway Museum commemorates the great railway engineer Timothy Hackworth (1786–1850), co-designer with William Hedley of the famous *Puffing Billy* (1813), now in the Science Museum, London. In 1825 he became responsible for the engineering shops of the Stockton and Darlington Railway. There he designed the powerful *Royal George* and the *Sans Pareil* which was entered, though unsuccessfully, in the famous Rainhill Trials in 1829.

SHORWELL, IoW

Yafford Mill, at Shorwell, is the only operational water-mill on the island. It dates from the 19th century and has been completely restored.

SHREWSBURY, Shrops.

Near the station at Ditherington is an old flax mill built by George Bage in the 1790s: it is noteworthy as being probably the first building ever constructed with a frame of cast iron, to avoid the risk of fire which had destroyed many of the early mills of the Industrial Revolution. It became a malthouse in 1886 and has since served various purposes.

Birthplace of Charles Darwin (1809–82), famous for his *Origin of Species* (1859). His father practised medicine there at The Mount and he went to Shrewsbury School before going to Edinburgh to study medicine. There is a bronze statue of him in front of the town library.

Clive House, so-called because it was the residence of Clive of India when he was Mayor in 1762, is now a museum and houses a fine collection of Coalport and Caughley porcelain.

SIBSEY, Lincs.

The six-sailed Sibsey Trader Mill, built in brick in 1877, has its machinery intact. There is a small associated museum.

SICCAR POINT, Loth. See COCKBURNSPATH

SIDMOUTH, Devon

Retirement home of Sir Norman Lockyer (1836–1920), famous for his discovery of helium in the Sun on spectrographic evidence. His private observatory – now restored after serious dilapidation – stands on Salcombe Hill as a lasting memorial.

SILBURY HILL, WILTS. See AVEBURY

SINGLETON, W Ssx. See CHICHESTER

SITES OF SPECIAL SCIENTIFIC INTEREST

By reason of their name alone the Sites of Special Scientific Interest cannot be omitted from a book such as this. The SSSIs are areas designated by English Nature (and the corresponding bodies in Scotland, Wales and Northern Ireland) as being of special scientific interest by reason of their flora, fauna, geological exposures, or other environmental singularities. Landowners/occupiers may be paid compensation in consideration of their desisting from certain normal field operations, such as early hay-making.

Many sites have been designated – currently over six thousand – and they range in size from less than one up to many thousands of hectares. Thus Westwood Quarry (Herts) is a mere tenth of a hectare while North Exmoor (Devon) extends to 12,000 hectares and the Dark Peak (W Yorks) to 32,000. Overall, sites in Britain total very nearly 2 million hectares, with a further 200,000 in Northern Ireland. Complete catalogues are published periodically by English Nature and the other regional bodies.

SKIDBY, Humb.

Skidby Windmill dates from 1821, and was operational using wind-power until 1954, when electric motors were installed. It is now in full working order and the associated buildings have been developed as a museum related to agriculture in general and milling in particular.

SKIPTON, N Yorks.

There has been a water-powered corn mill at Skipton since the 12th century and the present mill – part of the George Leatt Industrial and Folk Museum – is still operational. There is an associated collection of light horse-drawn vehicles.

The Yorkshire Dales Railway Museum (Embassy Rd) has a noteworthy collection of fully restored industrial locomotives.

SKREEN, Sligo

Birthplace of the physicist George Stokes (1819–1903), Lucasian Professor of Mathematics at CAMBRIDGE 1849–1903 and President of the Royal Society 1885–90. He is remembered for his work on the behaviour of fluids. Stokes' Law defines the resistance to motion of a sphere moving in a viscous liquid.

SLANE, Meath

Many regard the megalithic tomb of Newgrange (3 miles SE) as the finest example of its kind in the British Isles. The mound is 90 m in diameter and 14 m high. The stones at the entrance to the cruciform, stone-vaulted burial chamber are ornamented with carvings in the form of concentric rings and spirals. The chamber contains two stone bowls to receive cremated remains. That at Knowth (2 miles E) is second only to Newgrange. There the mound is 80 m in diameter and 12 m high. Two access passages are arranged back-to-back and lead to cruciform chambers, as at Newgrange.

SMETHWICK, W Mids.

The firm of W. & T. Avery, at Warley, is particularly identified with weights and weighing. The Avery Historical Museum has an important collection of items in this field, from ancient Egyptian and Roman times.

SNOWSHILL, Glos. See BROADWAY

SOLIHULL, W Mids

The National Motorcycle Museum displays several hundred British motor-cycles, dating from 1898.

SOUTHAMPTON, Hants.

A port since Roman times, it prospered in the Middle Ages and once again in the 20th century when it became Britain's leading port for passenger traffic. The first dock was built in 1842 but of this all that remains is the Princess Alexandra (outer) Dock. This business declined in the 1950s with the advent of jet airliners: in 1957 for the first time more passengers flew across the Atlantic than travelled by sea. The splendid Terminus Station, now used for other purposes, is a reminder of the great railway boom: it was built in 1839 by Sir William Tite (1798–1873) who also built the Royal Exchange in London.

Southampton has long been an important centre for the aircraft industry: over the years nearly thirty companies have been active in the area and this is reflected in exhibits in the Southampton Hall of Aviation (Albert Rd South). The numerous exhibits include the Spitfire, a Sandringham flying boat, helicopters and many related items.

The Southampton Maritime Museum (Town Quay) depicts another important aspect of the city's history. The exhibits include a number of historic marine steam engines.

Southampton University was founded in 1952 as part of the big university expansion programme in the 1950s.

SOUTH QUEENSFERRY, Loth.

The Forth Bridge, completed in 1890, is one of Britain's most outstanding industrial monuments. It was built to the design of John Fowler (1817–98) and Benjamin Baker (1840–1907). Earlier plans to build a bridge to the design of Thomas Bouch had been abandoned when his Tay Bridge collapsed disastrously in 1879 (see DUNDEE). It is 1½ miles long and has two 521 m spans and one of 210 m. Structurally, it has some resemblance to the Eiffel Tower completed in Paris by A.G. Eiffel a year earlier. But there is a very significant difference: the Eiffel Tower was the last major structure to be built of iron, the Forth Bridge the first to be built of steel.

SPALDING, Lincs.

The Pinchbeck Engine House displays a steam-engine and scoop dating from 1831, together with much other material relating to the draining of the Fens.

The Pode Hole Land Drainage Museum is situated in a pumping station built in 1826. The steam-pumping engine is no longer there but there is an interesting collection of exhibits relating to land drainage generally: a dragline excavator, oil and gas engines, and many other memorabilia.

SPARKFORD, Som.

The Haynes Sparkford Motor Museum has a large collection of cars and motor-cycles, including some rare marques.

SPITTAL, Dyfed

Scolton Manor (1840) is a fine country house, open to the public, with an associated museum of local history. A striking exhibit is the 0–6–0 locomotive *Margaret*, built in 1878 for the local Maenclochog and Rosebush Railway.

STAIGUE, Kerry

In eastern Ireland, where stone is plentiful, small fortified homesteads known as cashels were built. Typically, they consist of outer drystone walls – sometimes vitrified by fire – surrounding a central building surmounted by a fighting platform reached by flights of stairs. The cashel at Staigue is one of the finest in Ireland.

STALBRIDGE, Dors.

Stalbridge Manor was one of the estates of the 'great' Earl of Cork (1566–1643) and it was there that he sent his son, the Hon. Robert Boyle (1627–91) – 'the Father of Chemistry' – to study under a tutor after leaving Eton and before going on a Continental tour. On his father's death he inherited the estate and lived there 1645–55. It was there that chemistry – on which he left so distinctive a mark – became his consuming interest. Writing from there to his sister Lady Ranelagh in 1649 he said, 'Vulcan has so transported and bewitched me as to make me fancy my laboratory a kind of Elysium.'

STANLEY, Tays. See PERTH

STANMER, E Ssx

Stanmer Village Rural Museum is in the courtyard of Stanmer House, built in 1722 for the Earl of Chichester. It displays a variety of agricultural machinery and tools used by wheelwrights and blacksmiths. Less frequently encountered exhibits are a horse-gin and a donkey wheel.

STANTON DREW, Som.

A complex of stone circles located 9 miles W of BATH. It comprises three stone circles – one of which is the second largest in Britain – and two avenues. It was probably erected about 2500 BC.

STANWICK, N Yorks.

The large hillfort here was an important tribal centre for the Brigantes, and was extensively excavated by Sir Mortimer Wheeler (1890–1976). Many of the items he uncovered are displayed in the Yorkshire Museum, YORK.

STICKLEPATH, Devon

The Finch Foundry Museum comprises a restored edge-tool works with primitive water-powered machinery.

STOCKFIELD, Northld

The Hunday National Tractor and Farm Museum displays farm machinery, stationary engines and water-wheels.

STOCKTON-ON-TEES, Clev.

Famous in railway history as the terminus of the Stockton and DARLINGTON Railway, opened in 1825 using George Stephenson's *Locomotion* for traction.

Home of John Walker (1780–1859) who set up as an apothecary there about 1818. Remembered for his invention in 1827 of the first friction matches (lucifers).

STOKE BRUERNE, Northants.

The Waterways Museum has displays illustrating two centuries of canal history. Also examples of the colourful decoration of domestic items unique to the Boat People. (See also GLOUCESTER.)

STOKE-ON-TRENT, Staffs.

Industrial city on the River Trent world-famous for pottery manufacture. Immortalized as the Five Towns of Arnold Bennett's novels, now amalgamated as a single administrative unit. Industrial activity is represented in several local museums. The City Museum (Bethesda Street, Hanley) has a splendid collection of ceramics representative of local potteries from the 17th century onward. The Minton Museum has an extensive collection of fine Minton ware. The Sir Henry Doulton Gallery illustrates, with many fine examples, the history of Doulton ware since 1815. The Gladstone Pottery Museum, set in a restored factory at Longton, is more concerned with technical aspects and includes workshops, steam engines and bottle kilns. The Chatterley Whitfield Mining Museum is devoted to the history of the mining industry.

Josiah Wedgwood (1730–95) was born at Burslem and his famous Etruria Works was established nearby in 1769. Manufacture is now at Darlaston, where there is an impressive Wedgwood Museum and an associated craft centre.

The city is also the birthplace of Thomas Astbury FRS (1898–1961) who at the University of LEEDS did pioneer work on the application of X-rays to the analysis of polymers, especially fibres.

STOKESAY, Shrops.

Stokesay, a fortified manor house, has an impressive hall dating from the 13th century.

The massive roof timbers are an outstanding example of the woodworking skills of that time.

STONEHENGE , Wilts. See AMESBURY

STORNOWAY, Lewis, W Is.

The Isle of Lewis is rich in archaeological remains. Dun Carloway (15 miles N) is perhaps the best-preserved broch in Scotland, parts of the walls still standing to a height of over 6 m. Brochs were built over a relatively short period, roughly within the years 100 BC to AD 100. They were massive tower-like buildings, with walls up to 5 m thick at the base and rising to around 15 m; a stairway, serving various levels, was built within the wall thickness. The brochs were once believed to have been fortresses but it now seems that they were the homes of local chieftains.

The stone circle at CALLANISH (12 miles W) is one of the finest in Britain after STONEHENGE. A central stone 5 m high is surrounded by a circle of tall stones and from this radiates four rows of stones. Two other stones – Cnoc Fillibhir and Cnoc Ceann – are close by.

STOURBRIDGE, W Mids.

Famous for fine glassware since the early 17th century. This activity is reflected in the famous collection in the Broadfield House Glass Museum at KINGSWINFORD (6 miles N). More technical aspects of the history of glassmaking are demonstrated at Stuart Crystal's Red House Glassworks at Wordsley (4 miles N). (See also WEST BROMWICH.)

STOURPORT-ON-SEVERN, H & W

Located at the junction of the River Severn and the River Stour, Stourport began to be an important inland port in 1772 when the Staffordshire and Worcestershire Canal was opened, linking the Severn with the Trent and Mersey Canal and thus giving access to the Midland canal network. Numerous basins – still extant – were built to facilitate transfer of goods from river barges to canal narrow boats.

STOWMARKET, Sflk

The Museum of East Anglian Life occupies a 70-acre site at Abbot's Hall. The oldest feature is the Hall itself, with roof timbers dating from the 13th century; the largest is the former workshop of the engineering firm of Robert Boby, dating from 1870. There is also an 18th-century water-mill and a 19th-century windmill used for land drainage.

STRABANE, Tyr.

Birthplace of John Dunlap (1747–1812). Briefly, he worked at Gray's Printing Press (NT) in Main Street before emigrating to the USA. There he founded America's first daily newspaper, *The Pennsylvania Packet*, and printed the first broadside of the Declaration of Independence.

STRADBALLY, Laois

The Irish Steam Preservation Society displays a collection of standard and narrow-gauge locomotives at Stradbally Hall.

STRANGFORD, Down

Castle Ward Water Mill stands in the grounds of Castle Ward (NT). Adjacent is a Victorian laundry.

STRATFIELD SAYE, Berks.

Stratfield Saye was presented to the Duke of Wellington by a grateful nation after the defeat of Napoleon at Waterloo, and his descendants have lived there ever since. The National Dairy Museum, in Wellington County Park, is devoted to portraying the production, processing and marketing of milk.

STRATFORD-UPON-AVON, Warks.

The town is, of course, world-famous as the birthplace of William Shakespeare.

The Stratford and Moreton-in-Marsh Railway was completed in 1826 – as an extension of the Stratford Canal – only a year after the Stockton and Darlington Railway. An original wagon stands on a section of track near the canal basin.

A fine collection of firearms of all kinds, dating from the late 14th century, is displayed at the Arms and Armour Museum (Sheep Street), together with many other weapons and protective armour. The Stratford-upon-Avon Motor Museum (Shakespeare St) has a collection of cars mostly from the 1920s and 1930s. It includes a number of rare marques.

STREET, Som.

The Shoe Museum – located in the factory of C. & J. Clark (founded 1821) – exhibits shoes and related accessories from Roman times, as well as shoe-making machinery.

STRETTON, Ches.

Stretton Mill, water-powered, was built in the 16th century and over the years has been considerably modified. It was working until 1959 and has since been restored, but it is no longer operational.

STROMNESS, Ork.

The Ring of Brodgar is one of the largest and finest stone circles in Britain, built around 2500 BC. It is 110 m in diameter and twenty-seven of the original sixty stones survive. Nearby, and possibly once linked to it by a line of stones, is Stenness Henge, a much smaller circle: only four of the original twelve stones survive.

Skara Brae (5 miles N) has been called the 'British Pompeii'. It is a Neolithic village, the best preserved in Europe, which survived by being buried in sand during a storm. In 1850, another storm uncovered it.

STROUD, Glos.

From medieval times Cotswold wool has been prized as the finest in Europe and the production of cloth has long been an important local industry. With the Industrial Revolution, production became concentrated in mills, many originally powered by the River Frome which drains off the Cotswolds. One of the most impressive is at King's Stanley (3 miles W) built in 1813: it is an early example of an iron-framed building and in the manner of its day was designed in the classical style. As elsewhere, many mills have now been converted to other uses.

Stroud is the home of the mechanical lawn-mower invented by E.B. Budding in 1830 and first manufactured in the following year at the Phoenix Foundry. Ransome's of IPSWICH – now Europe's biggest manufacturer – acquired a licence to manufacture. Budding is said to have been inspired by the machines used to shear the nap from cloth in the local textile industry in which he worked. (See also NEWQUAY.)

STRUMPSHAW, Nflk. See NORWICH

STYAL, Ches.

Quarry Bank Mill was built in 1784 by the textile magnate Samuel Greg (1758–1834). Like many industrialists of his day, he associated the mill with a well-organized village for his employees. The mill is now a working museum of the cotton industry, illustrating a wide range of processes and machinery, including a large operational water-wheel dating from 1850.

SUNDERLAND, T & W

The Monkwearmouth Station Museum has a large transport exhibition including steam locomotives from the early 19th century to 1954. It is housed in the original 1848 railway station, now restored. The history of the glass industry, of great local importance, is illustrated in a section of Sunderland Museum and Art Gallery (Borough Rd).

Birthplace of the chemist Sir Joseph Swan (1828–1917), inventor of the incandescent filament lamp. He went to school there and for a time was apprenticed to a local pharmacist. (See NEWCASTLE UPON TYNE.)

SUNNINGWELL, Oxon.

Roger Bacon (1220–92), the famous OXFORD philosopher who dabbled in alchemy, is reputed to have come here to carry out some of his experiments. A notice in the church porch says:

> Our little village church (they say) was old
> When Roger Bacon loaned its gothic tower,
> To furnish Lady Science with a bower,
> Wherein to conjure Oxford's clay to gold.

SWAFFHAM, Nflk

The Iceni Village and Museum at Cockley Cley is a complex of four units. The imaginatively reconstructed Iceni village is believed to be on the original site. Additionally, there is a collection of horse-drawn vehicles dating back to the 19th century, a collection of agricultural machinery and equipment, and the East Anglian Museum, depicting the history of local life.

SWANSEA, W Glam.

Industrial city, particularly associated with coal, copper and tin. As a result of this activity – especially copper smelting – the Lower Swansea Valley became heavily polluted and virtually sterile. In recent years, however, a determined reclamation programme has reclaimed much of it. In the Clydach Gorge (7 miles N) are the remains of Clydach ironworks, including an iron tramway bridge of 1824. The Swansea and Mumbles Railway, opened in 1807, was the world's first passenger railway. It is now closed, but the line of the track remains. The first successful submarine telegraph cable was laid in Swansea Bay – between a boat and the Mumbles lighthouse – in 1844 by Charles Wheatstone (1802–75).

The Maritime and Industrial Museum (South Dock) includes a working woollen mill and a collection of historic vessels. The Royal Institution of South Wales Museum has exhibits relating to important local industries.

The famous NANTGARW pottery was established in 1813 by W. Billingsley and S. Walker, from WORCESTER.

SWINDON, Wilts.

For many years the Great Western Railway dominated Swindon's economy. The Great Western Museum displays GWR locomotives and rolling stock, and a wide range of railway memorabilia. At Wroughton (3 miles S) the Science Museum, London, houses much of its collection of agricultural machinery, road vehicles and commercial aircraft.

T

TACUMSHANE, Wex.

The thatched tower windmill on the north shore of Lake Tacumshane dates from 1846.

TALGARTH, Powys

Joseph Harris (1602–64) – mathematician, astronomer and Assay Master of the Mint – spent some time at Trevecca, home of his brother Howel, founder of a religio-industrial community. In 1761 he observed there a transit of Venus – an astronomically important event – and the telescope he used is in the Howel Harris Museum. He wrote important works on navigation and Edmond Halley (1656–1742) sent him on two voyages to the West Indies to test navigational instruments.

TANGMERE, W Ssx

Tangmere was an operational RAF airfield from 1916 until the end of the Second World War, during which it was of particular importance. The small museum illustrates the history of military flying in Britain from the earliest days.

TARRANT HINTON, Dors.

Many modern roads follow, but overlay, old ROMAN ROADS, so that exposures are rather rare. At Tarrant Hinton several miles of the raised agger of a Roman road crosses the A354, following the line of a right-of-way.

TATTERSHALL, Lincs.

The Dogdyke Pumping Station for land drainage dates from 1855 and contains a beam engine built in that year by Bradley and Craven of Wakefield. There is also an original 20-ft scoop wheel, the only one to have been worked by steam.

TAYNUILT, Strath.

The iron smeltery at Bonawe (25 miles from Oban on A85) was founded in 1753 to smelt Scottish iron ore with charcoal made in local forests. It continued in operation until 1876 and the works are well preserved.

TEALING, Tays.

Site of a well-preserved souterrain, lacking only the original stone roofing slabs (cf NEWBIGGING).

TEIGNMOUTH, Devon

Birthplace of Charles Babbage (1792–1871), pioneer of mechanical computing. His 'difference engine' was completed in 1834 and is now in the Science Museum, LONDON.

TELFORD, Shrops.

Coalbrookdale, where in 1709 iron was first successfully smelted with coke, rather than the traditional charcoal, by Abraham Darby (1678–1717) can fairly be regarded as the birthplace of the Industrial Revolution. The first iron bridge in the world (1779) still spans the River Severn there but is now open only to foot passengers. The Ironbridge Gorge Museum is a complex illustrating all aspects of local industrial history: the Museum of Iron and Furnace Site lies close to Darby's original blast furnace. Blist's Hill depicts life in a late Victorian industrial community. The Coalport China Museum is located on the site of the original Coalport factory, which closed in 1926. It has a fine collection of Coalport and Caughley china. The Jackfield Tile Museum is on the site of the old Craven Dunnill Tileworks. Natural oil was discovered in 1787 and exploited for many years. The original Tar Tunnel can be visited.

In 1987 UNESCO designated the Ironbridge Gorge a World Heritage Site.

TENBY, Dyfed

Birthplace of the mathematician Robert Recorde (*c.* 1510–58). He graduated at OXFORD in 1531 and was elected a Fellow of All Souls College. He subsequently had a chequered career in government service and ended his days in poverty, possibly as a debtor in King's Bench prison. Nevertheless he is remembered as the founder of the English school of mathematical writing and introduced the 'equal' (=) symbol into algebra.

TENTERDEN, Kent

The steam railway operates over five miles of track using vintage locomotives dating from the 19th century and seventy items of rolling stock. The original station buildings have been preserved.

The Kent and East Sussex Railway was the first to be constructed under the provisions of the Light Railways Act of 1896. An exhibition in the Tenterden and District Museum (Station Rd) is devoted to the history of light railways.

THETFORD, Nflk

At Weeting (7 miles NW) is Grimes Graves, where several hundred flint mines were opened around 2500 BC. Some shafts were as much as 10 m deep.

THORNHILL, W Yorks.

Home of the amateur astronomer John Michell (?1724–93) who was appointed Rector of St Michael's Church in 1767. He is remembered for two great achievements: the first realistic estimate of the distance of the stars and the discovery of double stars. A plaque on the church tower commemorates his work.

TILFORD, Sry

The Old Kiln Agricultural Museum illustrates the history of various aspects of local rural industry, particularly items relating to the growing and processing of hops. An associated exhibition exemplifies machinery – such as chain saws and pit saws – used in forestry and the timber industry.

TIVERTON, Devon

Home of John Heathcote (1783–1861) from 1816. In 1809 he invented a lace-making machine known as the bobbinet machine but his factory in LOUGHBOROUGH was destroyed by Luddites in 1816. He sought refuge in Tiverton, which he represented in Parliament 1832–59, and built a mill on the River Exe. The Tiverton Museum has a lace-machine gallery.

TODDINGTON, Glos.

The North Gloucestershire Railway Company (1985) has a large collection of steam locomotives and carriages, both standard and narrow gauge (both UK and foreign). There is also a fully restored signal-box from the old Midland Railway.

TONBRIDGE, Kent

The Milne Museum, maintained by the South-Eastern Electricity Board, has a large collection of electrical equipment 1800–1950. It includes both equipment – such as transformers and switchgear – and appliances, such as cookers and vacuum cleaners.

TORQUAY, Devon

Kent's Cavern is a show-cave which over the years has yielded many archaeological artefacts, some dating from Palaeolithic times. Many of them are now displayed in Torquay Museum and in the Natural History Museum, London.

TOTNES, Devon

The British Photographic Museum, Bowden House, displays a large collection of motion and still photographic cameras – some of rare marques – and related equipment.

Totnes Motor Museum (Steamer Quay) has a collection of vintage cars covering more than eighty years of motoring. Motor-bicycles and bicycles are also on display.

TREMADOC, Gynd

Town originally created in 1803–4 by the industrialist William Madocks (1773–1828) who built a woollen mill there. Later he reclaimed land on the Glaslyn Estuary and built a new harbour called Portmadoc. Here slate from quarries at Ffestiniog (see BLAENAU FFESTINIOG) was shipped. Later (1836) a narrow-gauge railway was built for this purpose, and this has now been restored by a preservation society as a tourist attraction.

Gwynedd Maritime Museum reflects the life of Portmadoc when it was a thriving seaport. A major exhibit is the *Garlandstone*, a 120-ton sailing ship built for the coastal trade in 1909.

TRESCO, IsS

The making of ships' figureheads is a traditional and highly specialized form of wood carving. A fine collection has been brought together at Valhalla, an outpost of the National Maritime Museum, Greenwich. The famous garden was started by Augustus Smith (1804–72).

TRING, Herts.

The Zoological Museum (British Museum (Natural History)) has a very large and well-described collection of mounted specimens of animals from all over the world.

TROWBRIDGE, Wilts.

Country town on the River Biss, long famous for woollen goods, especially Wiltshire broadcloth. Many of the mills built during the Industrial Revolution still survive. Stone Mill is interesting as it shows a transition from waterpower to steam: the remains of the water-wheel installations survive, as does the chimney of the steam-engine house.

The construction of Studley Mill, which spans the river, is unusual in that the walls are fenestrated with great numbers of small square apertures. This was to ensure free ventilation for drying the teasels used to raise the nap of the cloth.

TRURO, Corn.

The Cathedral, completed in 1910, contains a plaque commemorating the work of the Cornish astronomer John Couch Adams (1819–92), discoverer of the planet Neptune, who was born at LANEAST. Some of the stained-glass windows commemorate other scientists, including Newton and Francis Bacon.

TUAM, Gal.

The Mill Museum (Shop Street) is housed in a 300-year-old corn mill and has exhibits depicting the history of milling.

TYWYN, Gynd

The Tal-y-Llyn narrow-gauge railway was built to link the now long-closed slate mines at Abergynolwyn with Wharf Station at Tywyn. It is now once again a working railway, operating a regular service using some locomotives acquired from elsewhere. The Railway Museum at Tywyn includes a section devoted to the history of the local slate industry.

U

UCKFIELD, E Ssx

In 1912 the fossil skull of Piltdown Man was discovered in a gravel pit at Barkham Manor. He was hailed as *Eanthropus*, the Dawn Man, but in the 1950s was shown to be a deliberate hoax supposedly perpetrated by Charles Dawson, a local solicitor and archaeologist. The find consisted of a modern human cranium and the jawbone of an orang-utan. However, recent evidence suggests that Dawson was himself a victim, the site having been 'salted' with the remains for him to find.

The famous Bluebell Railway linking East Grinstead with Lewes was opened in 1882 and closed by British Railways in 1958. It was immediately taken over by a preservation society and was operational again in 1960, since when it has gathered together a considerable collection of steam locomotives and rolling stock. Sheffield Park Station has been maintained in 19th-century style and a museum of railway memorabilia has been set up on one of the platforms.

UFFCULME, Devon

Coldharbour Mill, dating from the 18th century, was originally used as a paper mill and then for grinding flour. In 1797 its use was changed again when it was acquired by a local woollen manufacturer, Thomas Fox, who needed extra capacity. Until 1978 power was supplied by an 18-ft breast-shot wheel, supplemented (1910–81) by a steam engine. The whole has now been turned into a working museum illustrating all the textile processes from spinning wool to weaving fabric.

UFFINGTON, Oxon.

Famous for the great hill-figure – traditionally a dragon but more likely a horse – delineated by cutting the turf to expose the white chalk. It measures 110 x 40 m. Possibly Iron Age – like the nearby Uffington Castle hillfort – but may even be Saxon.

Some 5 miles SW, at Ashbury, is Wayland's Smithy, a large stone burial chamber from which the overlaying soil has disappeared. Reputedly named after Volund the Smith, a Viking god who made shoes for the Uffington horse.

ULEY, Glos.

Hetty Pegler's Tump (1 mile N) is a fine example of a Neolithic long barrow. From the forecourt a passage 7 m long, with five chambers adjoining, leads into the mound.

UPMINSTER, Esx

William Derham FRS (1657–1735) was vicar of St Laurence, Upminster, 1689–1735. He published papers on meteorology, astronomy and natural history and edited works by Robert Hooke and John Ray. He is, however, best known for having made the first accurate measurement of the speed of sound. This he did by observing from the tower of his church the flash of a cannon fired at pre-arranged intervals at Blackheath, 20 km distant. This gave a value of 346 m/sec.: the accepted modern value under comparable atmospheric conditions is 341 m/sec.

USK, Gwent

Birthplace (Kensington House) of Alfred Russel Wallace (1823–1913), pioneer with Charles Darwin (1809–82) of the theory of evolution by natural selection, which they jointly presented to a meeting of the Linnaean Society in LONDON in 1858. Innately modest, he was subsequently content to be Darwin's disciple. Darwin once told him: 'You are the only man I have ever heard of who personally does himself an injustice, and never demands justice.'

VALENTIA ISLAND, Kerry

European terminus of the trans-Atlantic telegraph cable 1858–1966.

The lighthouse is of interest: until recent times it had been lit by acetylene gas, generated on the spot by the action of water on calcium carbide.

VENTNOR, IoW

The Blackgang Sawmill, housed in a 19th-century stone barn, is one of the few museums devoted to the timber industry. It includes a replica of a water-powered sawmill and illustrates the work of various craftsmen dependent on wood – notably carpenters, wheelwrights and coopers.

At St Catherine's Quay there is a reconstruction of a typical quayside scene as it would have appeared in Victorian times. Associated exhibits illustrate the work of the RNLI (see BARMOUTH), including a 1953 lifeboat.

W

WADHURST, E Ssx

Until deforestation created a shortage of charcoal, and the advent of smelting with coke caused a shift in the centres of industry, the Weald was the most important area in Britain for the production and working of iron. Production ceased early in the 19th century. An important manufacture was that of ornate firebacks: examples can be seen in the Museum of Local History at LEWES and the Museum and Art Gallery, HASTINGS.

A bizarre memorial to the industry is to be seen in the churchyard at Wadhurst, where there are more than thirty cast-iron grave slabs dating from 1617–1771.

WAKEFIELD, W Yorks.

The Yorkshire Mining Museum, based on the Caphouse Colliery at Overton, illustrates the history of coal mining from 1820, and includes underground galleries.

WALKERBURN, Bdrs

The Scottish Museum of Woollen Textile (Tweedvale Mill) is devoted to the history of the woollen industry.

WALLINGFORD, Berks.

Birthplace of Richard of Wallingford (c. 1291–1336), medieval astronomer and mathematician. Builder of one of the first astronomical clocks, at ST ALBANS, where he was appointed abbot in 1327. A description of the clock survives in the Bodleian Library, OXFORD.

WALSALL, W Mids

The Walsall Leather Centre, housed in a Victorian factory, has exhibits representing all aspects of the leather industry. (See also NORTHAMPTON.)

WANLOCKHEAD, Strath.

This village (on B797 10 miles NE of Sanquhar) is the centre of an area long worked for lead, silver and gold and reminders of this activity abound. One of the most striking, near Meadowfoot Cemetery, is a well-preserved water balance pump for mine drainage. The Museum of Scottish Lead Mining is in the centre of the village.

WANSDYKE

Britain's three great fortified barriers are the ANTONINE WALL, HADRIAN'S WALL and

OFFA'S DYKE. In addition, there are a number of shorter earthworks or dykes, of which the best-known is Wansdyke. This extended originally from Portishead (Som.) to Andover (Hants.) but it can now be clearly identified only in two sections, running roughly east-west between Dundry, near Bristol, and Great Bedwyn, near Marlborough (Wilts.). Its name is generally taken to signify Woden's Dyke and it dates from the Dark Ages. Its purpose is unclear, but it seems to have been designed to protect western and central Wessex against attacks from the north. Possibly, it simply marked the agreed boundary between Wessex and Mercia.

WARMINSTER, Wilts.

Longleat House, a popular tourist centre, has a small collection of scientific instruments begun by Lord Weymouth about 1700. It is housed in the Bishop Ken Library. Sir Christopher Wren (1632–1723), mathematician and architect, was born at East Knoyle (7 miles S).

WASHINGTON, T & W

The winding drum and steam engine of 'F' colliery have been preserved, and the engine-house turned into a museum of related coal-mining memorabilia.

WELLINGBOROUGH, Northants.

Irchester County Park is located within a spectacular old ironstone quarry. The Narrow Gauge Railway Museum displays locomotives and rolling stock used in the quarrying industry.

WELLOW, Som.

A notice at the entry to the chambered long barrow at Stoney Littleton describes it as 'the most perfect specimen of Celtic antiquity still existing in Great Britain'.

WELLS, Som.

After that at SALISBURY, the astronomical clock at Wells Cathedral (1392) is the oldest in Britain. Almost certainly both were made by the same hand – that of the Dutch clockmaker Johannes Lietuijt. The elaborate dial – one of the finest of its kind in the world – shows not only the hours but the phases of the moon and its attitude in the sky. The movement of the clock was replaced in 1885: the original is in the Science Museum, London.

WEST BROMWICH, W Mids

Home of James Keir FRS (1735–1820) from 1770. In 1771 he was appointed manager of a glassworks at STOURBRIDGE and ten years later founded the Tipton Chemical works nearby, supplying alkali to the local glass industry. This is generally regarded as the beginning of the scientific chemical industry in Britain.

WESTONBIRT, Glos.

Westonbirt Arboretum has one of the finest collections of mature trees and shrubs in Britain. Founded in 1829 by private landowners, it has since 1956 been the responsibility of the Forestry Commission. It now covers some 240 hectares, on which grow fifteen thousand trees and shrubs. It is both an important scientific collection and resource centre and a valuable public amenity, attracting around a quarter of a million visitors annually.

WESTON-SUPER-MARE, Avon

The International Helicopter Museum, Locking Airport, is the only one of its kind in Britain. It includes a number of rare items, including a Bristol-Belvedere, a Westland Widgeon and a Cierva C–30A.

The Woodspring Museum (Burlington Street) is devoted to local history, including industry. Displays illustrate the history of zinc and lead mining on Mendip since Roman times. There is also a large collection of cameras and other photographic material. The museum also houses items excavated at the local hillfort at Worlebury. There the line of the defences is evident above ground and there are a number of storage pits cut into the rock.

WESTONZOYLAND, Som.

The Pumping Station here was the first to be set up on the Somerset Levels. It displays steam engines, land-draining equipment, and a forge. There is also a narrow-gauge railway.

WEXFORD, Wex.

The Wexford Maritime Museum displays ship models and various items of maritime interest. It is also responsible for the lightship *Guillemot* in Wexford Harbour.

WHITBY, N Yorks.

Seaport and former whaling station. Its most famous citizen is Captain James Cook FRS (1728–79), famous as a navigator and for voyages of scientific exploration to the Pacific. The cottage at Great Ayton (20 miles W), where he spent his boyhood, has been dismantled and re-erected in Fitzroy Gardens, Melbourne. An obelisk marks the site. He learnt his seamanship as apprentice to a Whitby coal-shipper, John Walker, who lived in Grape Lane: the house is now the Captain Cook Memorial Museum. A statue of Cook stands on West Cliff, above the harbour.

Whitby is also closely identified with another scientific explorer, William Scoresby FRS (1789–1857), son of a whaling captain based on the port: he was born at Cropton (20 miles S). He made many scientific expeditions to Australia and the Arctic.

William Bateson FRS (1861–1926), pioneer of genetics, was born in Whitby. In 1900 he drew attention to Gregor Mendel's long-forgotten work on inheritance in peas carried out in the Augustinian Monastery at Brno (Czech Republic).

At Ugthorpe (6 miles W) is Loose Howe, a long barrow dated at 1700 BC. It is remarkable because it was found to contain a body in a tree-trunk coffin together with a dugout canoe.

WHITWELL, Derbs.

Creswell Crags has been designated a Site of Special Scientific Interest. In the narrow gorge some twenty caves were occupied by Neanderthal man from around 40,000 BC.

WIDNES, Ches.

The Museum of the Chemical Industry (Catalyst Museum, Mersey Road) is devoted to the chemical industry and its social impact.

Birthplace of Charles Glover Barkla (1877–1944), physicist noted for research on X-rays.

Home from 1857 until his death of the Swiss industrial chemist Ferdinand Hurter (1844–98). He was closely identified with the locally very important alkali industry and, when the United Alkali Co. was formed in 1890, he set up the first industrial research laboratory in Britain. An older generation of photographers will remember him for the old 'H & D' rating for film speed, with the D referring to his friend V.C. Driffield (1848–1915).

WIGAN, GM

Wigan Pier Museum illustrates local life and industry. Exhibits include a steam-driven mill, colliery and rope-making machinery.

At Trencherfield Mill there is a large steam engine constructed in 1907.

WIGTON, Cumb.

The Nobel Laureate Sir William Bragg (1862–1942), famous for his use of X-rays to determine the arrangement of atoms in crystals, was born at Stoneraise Place, Westward. There is a commemorative plaque over the front door of this remote farmhouse.

WILLENHALL, W Mids

Willenhall is the main centre of the British lock industry, associated with such famous names as Union, Squire and Yale. The Lock Museum (Walsall St) displays many fine examples of the locksmith's craft, old and new, and a model of a traditional workshop.

WILTON, Wilts.

A town long famous for fine carpets, and the site of the oldest carpet factory in Britain. The Wilton Royal Carpet Factory has a display of carpet-making crafts.

WIMBORNE MINSTER, Dors.

The parish church has a fine astronomical clock dating from the 17th century. It illustrates the medieval concept of the Universe: the Earth is at the centre of the dial and the Sun revolves round it to show the hours.

WINTERBORNE ABBAS, Dors.

The Nine Stones is a small stone circle dating from the 2nd millennium BC.

WIRKSWORTH, Derbs.

The Middleton Top Engine House houses a fully restored beam engine built in 1830: it is now worked by compressed air. Two huge 14-ft winding wheels also survive. The machinery was used to haul loaded trucks up an incline on the Cromford and High Peak Railway.

Home of Abraham Bennet (1749–99), pioneer in the experimental study of electricity. He was ordained in London and appointed curate of Wirksworth in 1776, an office which he held until his death. The publication of his *New Experiments in Electricity* in 1789 led to his election to Fellowship of the Royal Society. His most important invention was the gold-leaf electroscope, still in use 200 years later.

WITNEY, Oxon.

Witney, a market town on the River Windrush, has long been a centre of the Cotswold woollen industry, particularly the making of blankets. Erly's factory dominates the centre of the town.

The Cogges Farm Museum, built round a 13th-century manor house, illustrates life on a large Oxfordshire farm at the beginning of this century. There are regular demonstrations of rural crafts such as hurdle-making, thatching and horse-shoeing.

WOLVERHAMPTON, W Mids

Manufacturing town in the heart of the Black Country. In the 1770s it became the centre of the japanning industry (see PONTYPOOL) in which trays and other domestic items were fashioned in papier mâché or layered paper, coated with bitumen, and then stoved to give a hard black finish. The centre of the industry was Old Hall, and it flourished until the 1850s. Examples can be seen in Bantock House Museum.

The Aerospace Museum (Cosford), located on a RAF airfield, has one of the finest civil and military aviation collections in Britain. Aircraft displayed include the Bristol Freighter, the Hastings, and the Vulcan and Victor bombers.

Birthplace of Sir William Bayliss FRS (1860–1924), famous for his research on the physiology of the digestive, vascular and nervous systems. He is particularly noted for his discovery, in 1902, of the hormone secretin which stimulates the pancreas to produce insulin.

WOODHENGE , Wilts. See AMESBURY

WOODSTOCK, Oxon.

Blenheim sawmills – on the Duke of Marlborough's estate at Combe – preserves much of the machinery of the mid-19th century mill. This includes a working beam-engine, Cornish boiler and a blacksmith's forge.

WOOL, Dors.

The Tank Museum at Bovington Camp displays some two hundred armoured fighting vehicles, dating from 1915.

WOOLPIT, Sflk

The Woolpit Bygones Museum has an exhibit devoted to the local brick-making industry, which has been practised for some four hundred years.

WORCESTER, H & W

The Dyson Perrins Museum at Worcester Royal Porcelain Works houses the world's finest and most comprehensive collection of old Worcester porcelain.

The Dyson Perrins family became famous as manufacturers of Lea and Perrins Worcester Sauce. They also became rich and made many large benefactions, among them a large contribution to the Dyson Perrins Chemical Laboratory in OXFORD, where the first professor was W.H. Perkin Jr (1860–1929), son of W.H. Perkin (1838–1907) discoverer of mauvein and thus founder of the synthetic dyestuffs industry.

WORSBROUGH, S Yorks.

The two mills at Worsbrough are both operational and regularly grind flour. One dates from the 17th century and is water-powered; the other, 19th century, has been converted to drive by an oil engine. The two are the centre of a museum complex.

WORTHING, W Ssx

At Cissbury some two hundred shallow pits mark one of the largest groups of flint mines in Neolithic England. Some of the shafts were 12 m deep, and many lie within the ramparts of a hillfort of later (Iron Age) date. (See THETFORD.)

WORTLEY, S Yorks. See SHEFFIELD

WOTTON-UNDER-EDGE, Glos.

Birthplace of the chemist Sir Charles Blagden FRS (1748–1820). He knew many British and French chemists of his day, and it was through him that Lavoisier learnt of Henry Cavendish's important experiments on 'inflammable air' (hydrogen). Cavendish was a wealthy man – 'the richest of the learned and the most learned of the rich' – and left Blagden £15,000 and an annuity of £500. After Lavoisier's execution in 1794, Blagden unsuccessfully courted his widow.

Blagden made important, and very accurate, measurements of the freezing-points of solutions, and is remembered by Blagden's Law, according to which the depression of the freezing-point is proportional to the concentration of the solution.

WREXHAM, Clwyd

The Geological Museum of North Wales is located in Bwlchgwyn Quarry and displays rocks and minerals from all over the world. There is a complementary display of industrial equipment.

WYE, Kent

The Agricultural Museum of Wye College, which is located in a 14th-century barn and a 19th-century oasthouse, has a good display of agricultural machinery, mostly horse-powered.

WYLAM, Northld

The Wylam Waggonway is famous as the track on which the locomotives *Wylam Dilly* and *Puffing Billy* ran. The Wylam Railway Museum displays railway memorabilia, photographs and other items. Close by is the birthplace (NT) of George Stephenson (1781–1848).

YZ

YELVERTON, Devon

Dartmoor is rich in megalithic remains, of which Down Tor (4 miles ENE) is one of the most interesting. A row of more than 150 stones – culminating in a massive 5-ton pillar – leads up to a circle of thirty stones.

YEOVIL, Som.

The Fleet Air Arm Museum, Yeovilton, depicts the development of naval aviation and displays more than fifty aircraft. The adjacent *Concorde* Exhibition allows visitors to see the interior of the first (experimental) *Concorde* airliner.

At South Cadbury (10 miles NW) is Cadbury hillfort, occupied from about 3000 BC: traditionally it was the site of Arthur's Camelot.

YORK, N Yorks.

Founded AD 71 as capital of the Roman province of Britannia and captured by the Danes in AD 876. The Jorvik Viking Centre vividly depicts life in the city during the Viking period.

The 14th-century Bellfounders Window in York Minster illustrates in detail the technique of bellfounding at that date. A window in All Saints, North Street, depicts the last fifteen days of the world, an interesting example of medieval cosmology.

The Castle Museum has a fine collection of folk material. Much of it is organized as reconstructed workshops, including that of T. Cooke (1807–68), a famous maker of clocks and scientific instruments. It also has one of the finest collections of scientific instruments outside the big national collections.

The National Railway Museum (an outpost of the Science Museum, London) houses the most important collection of its kind in Britain. It includes famous locomotives and rolling stock and a vast range of railway memorabilia. The University has lately founded an Institute of Railway Studies.

The city is famous for its chocolate industry, founded by the Quaker Rowntree brothers, Joseph and Isaac, at Tanners Mount on the banks of the River Ouse in 1869. It was barely profitable until 1879, when a Frenchman, Claude Gaget, visited them with samples of gums and pastilles, hitherto a French monopoly. Thereafter the business flourished and in 1898 a new factory, equipped with the latest chocolate processing plant, was opened on Haxby Road. By the turn of the century Rowntree's were rivalling the two other great Quaker chocolate manufacturers – Cadbury's of BIRMINGHAM and Fry's of BRISTOL.

York was the home from 1781 until the end of his short life of John Goodricke (1764–1786), the astronomical prodigy who, despite being a deaf-mute, has an important place in the history of science. In 1781 he made the then startling observation that the star beta-Persei (Algol) fluctuated regularly with a periodicity of about two days. He measured the periodicity carefully and obtained a value very close to that now accepted. He also correctly deduced the cause of the variation: 'It could hardly be accounted for other than . . . by the interposition of a large body revolving around Algol.' In modern parlance, Algol is a binary star.

In 1644 the Royalists suffered a disastrous defeat at the Battle of Marston Moor. Among the dead was the young astronomer William Gascoigne (c. 1612–44) who made two important improvements to the telescope. One was the introduction of cross hairs in the focal plane to allow accurate alignment: the other was the invention of a micrometer to measure small angular distances.

York was the site of the inaugural meeting of the British Association for the Advancement of Science in 1831. It was organized by William Vernon Harcourt FRS (1789–1871) – son of Edward Harcourt, then Archbishop of York – who combined his duties as vicar of Bishopsthorpe with studying chemistry in his private laboratory. The BAAS – which anybody interested in science could join – was founded to counterbalance the Royal Society of LONDON, which many scientists thought had become too professional and restricted.

ZENNOR, Corn.

The Wayside Museum is based on an old mill complex. The mill building contains original machinery and a blacksmith's forge. There is also a collection of mining and quarrying tools and equipment.

MAPS

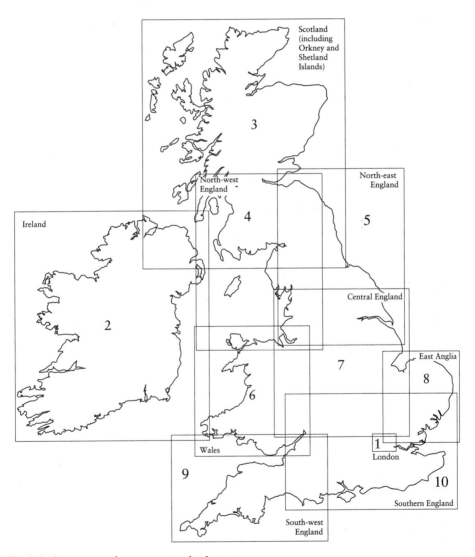

Scotland
(including
Orkney and
Shetland
Islands)

3

North-west
England

Ireland

4

North-east
England

5

Central England

2

East Anglia

7

8

6

Wales

1

London

9

10

Southern England

South-west
England

Shaded place names have an entry in the text

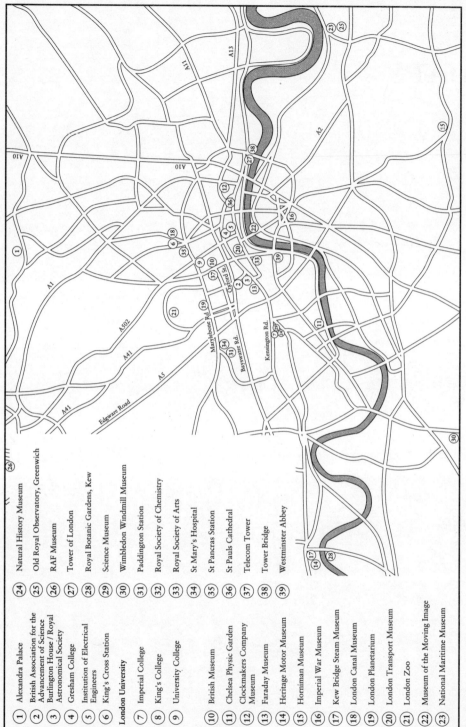

1. London

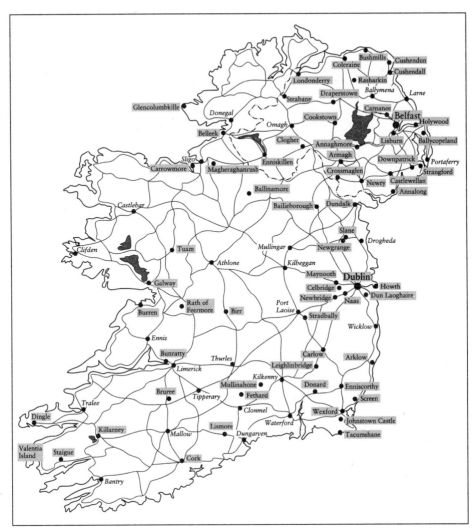

2. Ireland

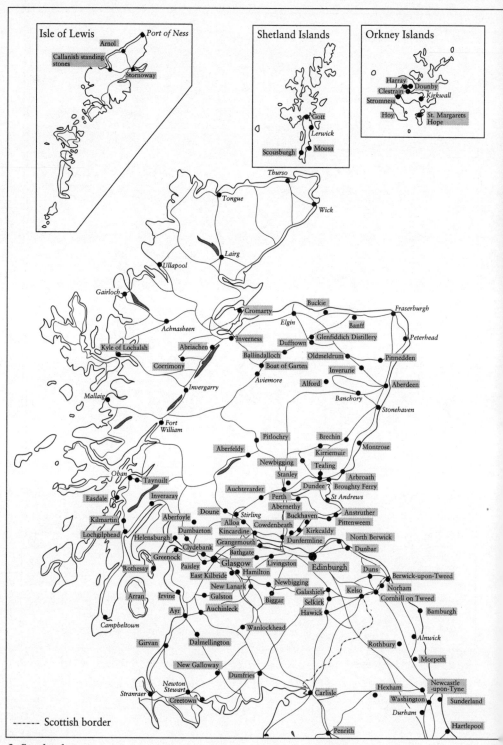

Isle of Lewis

Port of Ness

Arnol

Callanish standing stones

Stornoway

Shetland Islands

Gott

Lerwick

Scousburgh   Mousa

Orkney Islands

Harray   Dounby

Clestrain   Kirkwall

Stromness

Hoy   St. Margarets Hope

Thurso

Tongue   Wick

Lairg

Ullapool

Gairloch

Achnasheen

Cromarty   Buckie   Fraserburgh

Elgin   Banff

Inverness   Glenfiddich Distillery   Peterhead

Kyle of Lochalsh   Abriachen   Dufftown

Ballindalloch   Oldmeldrum   Pitmedden

Corrimony   Boat of Garten   Inverurie

Aviemore   Alford   Aberdeen

Invergarry

Mallaig   Banchory   Stonehaven

Fort William

Pitlochry   Brechin

Aberfeldy   Kirriemuir   Montrose

Newbigging   Tealing

Stanley   Arbroath

Oban   Taynuilt   Auchterarder   Dundee   Broughty Ferry

Easdale   Perth   St Andrews

Inveraray   Abernethy

Kilmartin   Doune   Stirling   Buckhaven   Anstruther

Aberfoyle   Alloa   Cowdenbeath   Pittenweem

Lochgilphead   Dumbarton   Kincardine   Kirkcaldy

Helensburgh   Grangemouth   Dunfermline   North Berwick

Clydebank   Bathgate   Dunbar

Greenock   Glasgow   Livingston

Rothesay   Paisley   Edinburgh   Duns

Arran   East Kilbride   Hamilton   Berwick-upon-Tweed

Irvine   New Lanark   Newbigging   Kelso   Norham

Galston   Galashiels   Cornhill on Tweed

Ayr   Auchinleck   Biggar   Selkirk   Bamburgh

Hawick

Wanlockhead

Campbeltown   Rothbury   Alnwick

Girvan   Dalmellington   Morpeth

New Galloway

Dumfries   Hexham   Newcastle-upon-Tyne

Newton Stewart   Washington   Sunderland

Stranraer   Creetown   Carlisle   Durham

Penrith   Hartlepool

------ Scottish border

3. Scotland

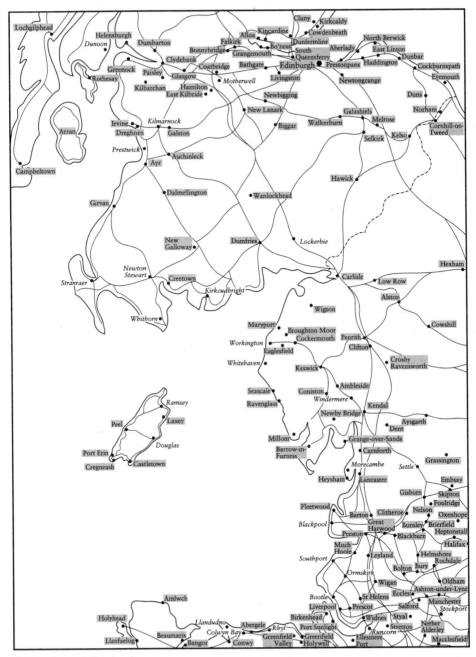

4. North-west England

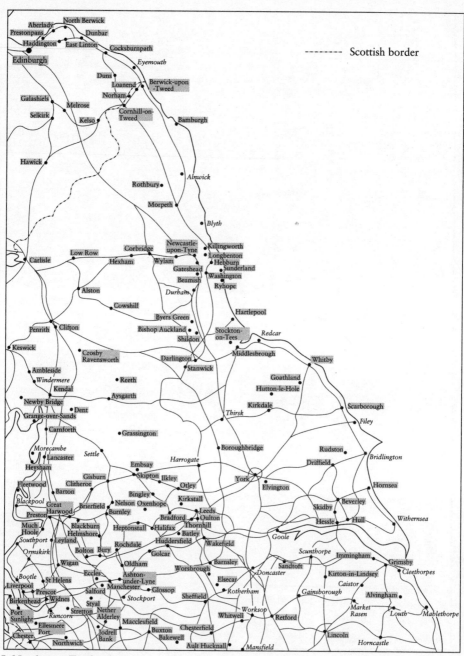

---------- Scottish border

Aberlady
North Berwick
Prestonpans
Dunbar
Haddington
East Linton
Cocksburnpath
Edinburgh
Eyemouth

Duns
Loanend
Berwick-upon-
-Tweed
Norham
Galashiels
Melrose
Cornhill-on-
Selkirk
Kelso
Tweed
Bamburgh

Hawick

Rothbury
Alnwick

Morpeth

Blyth

Corbridge
Newcastle-
upon-Tyne
Killingworth
Carlisle
Low Row
Longbenton
Hexham
Wylam
Hebburn
Gateshead
Sunderland
Beamish
Washington
Durham
Ryhope

Alston

Cowshill
Hartlepool
Byers Green
Penrith
Clifton
Bishop Auckland
Stockton-
Redcar
Keswick
Shildon
on-Tees
Crosby
Ravensworth
Darlington
Middlesbrough
Whitby
Ambleside
Stanwick
Windermere
Goathland
Reeth
Hutton-le-Hole
Kendal
Newby Bridge
Aysgarth
Kirkdale
Scarborough
Dent
Grange-over-Sands
Thirsk
Filey
Carnforth
Grassington

Morecambe
Boroughbridge
Rudston
Lancaster
Settle
Bridlington
Heysham
Embsay
Harrogate
Driffield
Fleetwood
Gisburn
Skipton
Ilkley
York
Hornsea
Clitheroe
Blackpool
Barton
Bingley
Otley
Elvington
Great
Nelson Oxenhope
Kirkstall
Skidby
Beverley
Preston
Harwood
Brierfield
Burnley
Leeds
Hessle
Hull
Much
Blackburn
Bradford
Oulton
Withernsea
Hoole
Helmshore
Heptonstall
Halifax
Thornhill
Southport
Leyland
Batley
Goole
Ormskirk
Bolton
Bury
Rochdale
Huddersfield
Wakefield
Wigan
Golcar
Scunthorpe
Immingham
Grimsby
Oldham
Barnsley
Sandtoft
Bootle
Eccles
Worsbrough
Doncaster
Kirton-in-Lindsey
Cleethorpes
Liverpool
St Helens
Ashton-
Elsecar
Caistor
Prescot
Salford
under-Lyne
Glossop
Rotherham
Gainsborough
Alvingham
Birkenhead
Widnes
Styal
Stockport
Sheffield
Worksop
*Market
Port
Runcorn
Nether
Whitwell
Retford
Rasen*
Louth
Mablethorpe
Sunlight
Stretton
Alderley
Macclesfield
Chesterfield
Lincoln
Ellesmere
Jodrell
Buxton
Chester
Port
Bank
Bakewell
Horncastle
Northwich
Ault Hucknall
Mansfield

5. North-east England

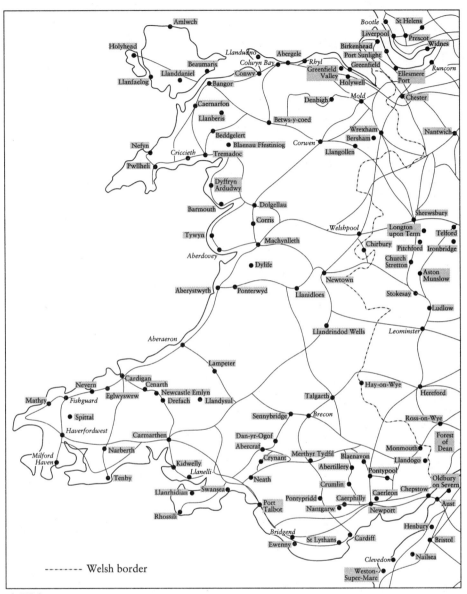

6. Wales

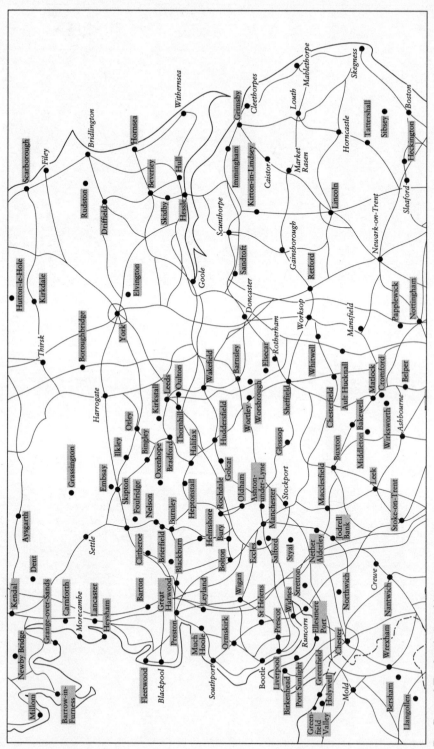

7a. Central England (north)

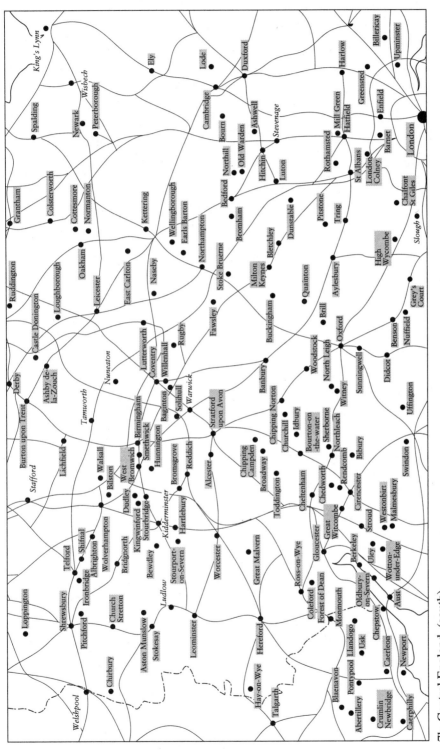

7b. Central England (south)

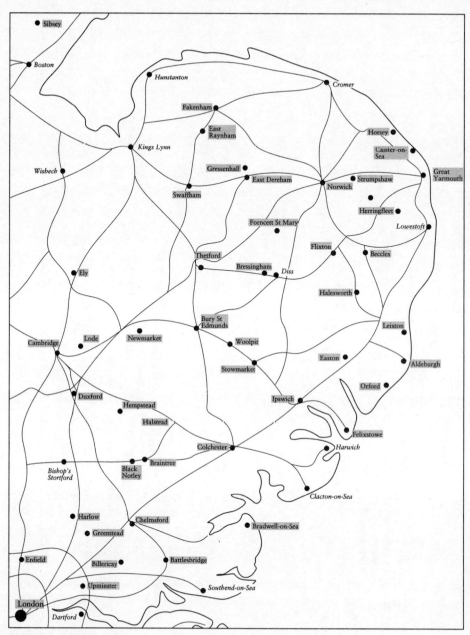

Sibsey

Boston

Hunstanton

Cromer

Fakenham

East
Raynham

Horsey

Kings Lynn

Caister-on-
Sea

Wisbech

Gressenhall

Great
Yarmouth

East Dereham

Norwich

Strumpshaw

Swaffham

Herringfleet

Forncett St Mary

Lowestoft

Thetford

Flixton

Beccles

Bressingham

Diss

Ely

Halesworth

Bury St
Edmunds

Leiston

Lode

Newmarket

Woolpit

Cambridge

Easton

Aldeburgh

Stowmarket

Duxford

Orford

Hempstead

Ipswich

Halstead

Felixstowe

Colchester

Harwich

Black
Notley

Braintree

Bishop's
Stortford

Clacton-on-Sea

Harlow

Chelmsford

Greenstead

Bradwell-on-Sea

Enfield

Battlesbridge

Billericay

Upminster

Southend-on-Sea

London

Dartford

8. East Anglia

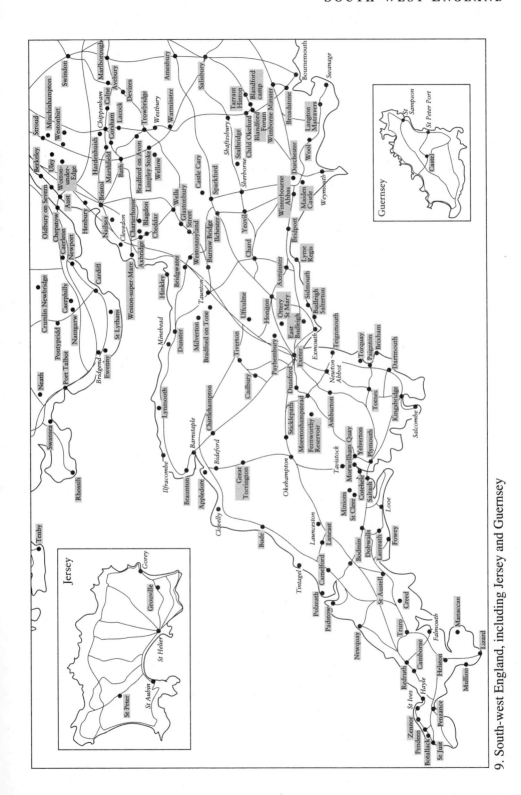

9. South-west England, including Jersey and Guernsey

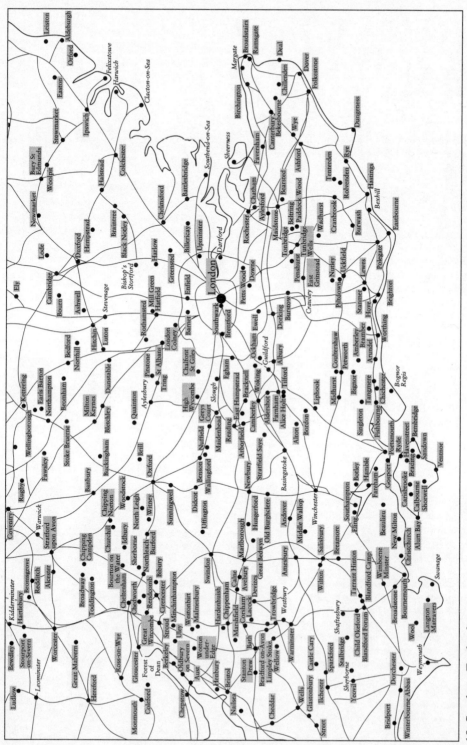

10. Southern England

DRAMATIS PERSONAE
Principal Scientists, Engineers and Others

Ackermann, Rudolf (1764–1834)
Inventor of the Ackermann steering system for cornering vehicles. His print shop in the Strand, London, was one of the earliest premises to be illuminated by gas.
see LONDON

Adams, John Couch (1819–92)
Astronomer. In 1843 deduced the position of Neptune, an unknown planet, but through the dilatoriness of Sir George Airy (q.v.), Astronomer Royal, credit went to the French astronomer Urbain Le Verrier.
see CAMBRIDGE, LANEAST, TRURO

Adamson, Daniel (1820–90)
Ironfounder: chief promoter of the Manchester Ship Canal.
see MANCHESTER

Adamson, Robert (1821–48)
Early professional photographer who in the 1840s established a fashionable studio in Edinburgh in collaboration with the painter David Octavius Hill.
see EDINBURGH

Addison, Thomas (1793–1860)
Physician. In 1855 he described the characteristic symptoms of what is now called Addison's disease, a disorder of the suprarenal capsules. Regarded as the founder of endocrinology.
see LANERCOST, LONGBENTON

Adrian, Edgar (Baron Adrian of Cambridge 1889–1977)
Physiologist renowned for his highly original research on the conduction of nervous impulses, for which he shared a Nobel Prize in 1932 with Charles Sherrington (q.v.).
see CAMBRIDGE

Airy, George (1801–92)
Astronomer Royal 1835–81, during which time Britain adopted Greenwich Mean Time as legal standard (1880). Contributed to planetary theory:

led British expedition to observe a transit of Venus in 1874.
see IPSWICH

Alcock, John William (1892–1919)
Pioneer aviator: made first non-stop trans-Atlantic flight with Arthur Whitten Brown (q.v.) in 1919.
see MANCHESTER

Allen, John (1670–1741)
Physician whose textbook of medicine was widely used in Britain and on the Continent. Versatile inventor of mechanical devices.
see BRIDGWATER

Anderson, Adam (d.1846)
Physicist: one-time Rector of St Andrew's University. Pioneer of building construction in cast-iron.
see PERTH

Anderson, John (1726–94)
While Professor of Natural Philosophy in Glasgow University he began a science class for working men. Left his fortune to endow Anderson's College, now Strathclyde University.
see GLASGOW

Andrews, Thomas (1813–85)
Chemist who demonstrated the continuity of the liquid and vapour states and paved the way to the liquefaction of the so-called permanent gases.
see BELFAST

Anning, Mary (1799–1847)
Keen amateur geologist. Remembered for having in 1811 discovered complete skeleton of icthyosaurus, and later plesiosaurus and pterodactylus.
see LYME REGIS

Appleton, Edward (1892–1965)
Physicist, remembered for his research on the upper atmosphere, where he identified an ionized reflecting layer (Appleton Layer) important for radio transmission. Nobel Laureate 1947.
see EDINBURGH, BRADFORD

Arkwright, Richard (1732–92)

Pioneer of the mechanical processing of textile fibres and rightly regarded as a founder of the Industrial Revolution. Built industrial complexes in which factories were integrated with workers' villages.
see CROMFORD, LONDON, NEW LANARK, PERTH, PRESTON

Armstrong, William (1810–1900)

Engineer, with particular interest in armaments. He saw the industrial possibilities of electricity and his own home was the first house in the world to be so illuminated.
see ROTHBURY

Ashmole, Elias (1617–92)

Antiquary/scientist. Author of Theatrum Chymicum (1652) and founder of the Ashmolean Museum, Oxford.
see OXFORD

Astbury, William (1898–1961)

Pioneer of application of X-rays to analysis of the structures of polymers, especially textile fibres.
see LEEDS, STOKE-ON-TRENT

Aston, Francis (1877–1945)

Pioneer atomic physicist, remembered particularly for his mass spectrograph, which separated particles of different masses much as the optical spectrograph separates light of different wavelengths (colours). Nobel Laureate 1922.
see CAMBRIDGE

Babbage, Charles (1792–1877)

Lucasian Professor of Mathematics, Cambridge, 1828–39. Pioneer of mechanical computing. Lost money seeking to devise infallible system for forecasting results of horse races.
see LONDON, TEIGNMOUTH

Bacon, Anthony (1718–86)

Welsh ironmaster, whose memorial is Cyfarthfa Castle, now an industrial museum.
see MERTHYR TYDFIL

Bacon, Francis (Viscount St Albans 1561–1626)

Statesman/scientist. Held high office under James I but convicted of corruption 1621. Conceived a grandiose scheme of embodying the whole of human knowledge in a Great Instauration which would benefit mankind as a whole.
see CAMBRIDGE, LONDON, ST ALBANS

Bacon, Roger (1214–92)

Great medieval scholar (Doctor mirabilis) often regarded as first scientist in modern sense.
see ILCHESTER, OXFORD, SUNNINGWELL

Bain, Alexander (1810–77)

Effected improvements in electric telegraph and constructed electric clocks.
see EDINBURGH

Baird, John Logie (1888–1946)

Pioneer of practical television, but his electromechanical system of transmission and reception had to give way to the modern all-electronic system in the 1930s.
see HELENSBURGH, LONDON

Baker, Benjamin (1840–1907)

Civil engineer. Designer, with John Fowler (q.v.), of the Forth Bridge (1890).
see IDBURY, SOUTH QUEENSFERRY

Bakewell, Robert (1725–95)

Pioneer of scientific agriculture, famous for breeding improved strains of farm animals.
see LOUGHBOROUGH

Balfour, Andrew (1630–94)

Physician: founder of the Edinburgh Physic Garden (1670).
see EDINBURGH

Ball, Robert (1840–1913)

Astronomer: served as Astronomer for Ireland 1874–92 before going to Cambridge as Professor of Astronomy and Geometry.
see DUBLIN

Banks, Joseph (1743–1820)

Botanist: President of the Royal Society for 42 years from 1778 and undoubtedly the most influential scientist of his day. Accompanied James Cook (q.v.) as naturalist on the Endeavour expedition (1768–71) to observe a transit of Venus and explore the Pacific Ocean. Played a leading role in establishing Kew Gardens, and also the first Australian colony at Botany Bay.
see LINCOLN, LONDON

Barcroft, Joseph (1872–1947)

Physiologist, famous for his research on the transportation of oxygen in the blood and on respiration generally.
see NEWRY

Barrow, Isaac (1630–77)

Mathematician and classical scholar. In 1669 he generously resigned his chair of mathematics at Cambridge in favour of Isaac Newton (q.v.) whose ability he recognized as superior to his own.
see CAMBRIDGE, LONDON

Bartlett, Neil (b.1932)

Chemist who in 1962 prepared the first compounds of the inert gases, until then supposed to be wholly lacking in chemical reactivity.
see NEWCASTLE UPON TYNE

Baskerville, John (1706–75)

Printer and type-founder, famous for his design of individual characters.
see BIRMINGHAM

Bateman, John Frederic (1810–89)

Civil engineer, concerned with many major water storage and distribution projects in the British Isles.
see ABERFOYLE

Bates, Henry Walter (1825–92)

Naturalist and Amazonian explorer. He collected eight thousand species of hitherto unknown insects.
see LEICESTER

Bateson, William (1861–1926)

Pioneer of genetics: brought to public attention the forgotten experiments carried out by Gregor Mendel in the 1860s.
see WHITBY

Bayliss, William (1860–1924)

Physiologist, remembered particularly for his discovery in 1902 of the hormone secretin, which regulates the secretion of the pancreas.
see WOLVERHAMPTON

Bazalgette, Joseph (1819–91)

Civil engineer, remembered for his construction (1855–89) of a complete new drainage system for London, involving 100 miles of large-diameter sewers. This was prompted by two major outbreaks of cholera. The scheme involved a new Thames embankment between Westminster and Blackfriars Bridges.
see EDINBURGH, LONDON

Beddoes, Thomas (1760–1808)

Chemist, remembered for his work on the medical uses of gases, which he pursued at his Medical Pneumatic Institute in Bristol. For a time Humphry Davy (q.v.) was his assistant there.
see BRISTOL, SHIFNAL

Bell, Alexander Graham (1847–1922)

An early interest in sound waves, and the mechanics of speech, led him to improve the electric telegraph and eventually to the invention of the telephone.
see EDINBURGH

Bell, Patrick (1799–1869)

Born into a Scottish farming family, Bell became interested in the development of a mechanical shear to supersede his father's reaper. This led to a practical machine in 1827. Years later the importance of this invention was recognized by an award of £1000.
see DUNDEE

Bennet, Abraham (1749–99)

Pioneer of the study of electricity by practical experiments.
see WIRKSWORTH

Bentham, George (1800–84)

Plant taxonomist. In 1854 he presented his valuable herbarium to Kew Gardens, where he worked for the rest of his life. His popular *Handbook of the British Flora* (1858) went into several editions.
see LONDON

Bessemer, Henry (1813–98)

Steel manufacturer famous for the Bessemer Process for decarburizing iron. Apart from this, an extraordinarily fertile inventor in many other fields.
see HITCHIN

Bickford, William (1774–1834)

His invention of the safety fuse for explosives (1831) ranks in importance with Davy's invention of the miners' safety lamp.
see CAMBORNE

Biffen, Rowland (1874–1949)

Geneticist: appointed first Director of the Plant Breeding Station, Cambridge, in 1912. Contributed to the development of rust-resistant wheat.
see CHELTENHAM

Birkbeck, George (1776–1841)

Physician: began his career as a lecturer at Anderson's College, Glasgow. Moved to London in 1804 and founded the London Mechanics' Institute there, now Birkbeck College, London University.
see GLASGOW

Black, James (b.1924)

Pharmacologist. After academic appointments he embarked on a career in the pharmaceuticals industry, initially with ICI. Discovered the important class of drugs known as beta-blockers and introduced cimetidine for treatment of stomach ulcers. Nobel Laureate 1988.
see COWDENBEATH

Black, Joseph (1728–99)

Chemist: did pioneer research on the chemistry of gases and formulated the concepts of specific and latent heat.
see GLASGOW

Blackett, Patrick (Baron Blackett 1897–1974)

As a physicist Blackett is remembered particularly for his research on cosmic rays and the magnetism of celestial bodies. During the Second World War he had a distinguished career in operational research, and afterwards had a powerful influence on scientific and technological policy in Britain. Nobel Laureate 1948.
see MANCHESTER

Blackwell, Elizabeth (1821–1910)

First woman to be placed on the British Medical Register (1859).
see BRISTOL

Blagden, Charles (1748–1820)

Chemist, remembered for Blagden's Law relating the freezing-point of a solution to its concentration. Became involved in the great 'Water Controversy' between Henry Cavendish (q.v.) and Lavoisier on the chemical composition of water. Unsuccessfully courted Lavoisier's widow.
see WOTTON-UNDER-EDGE

Bodley, Thomas (1545–1613)

Scholar and diplomat and one of the great benefactors of Oxford University, notably in founding the world-famous Bodleian Library.
see OXFORD

Bohr, Niels(1885–1962)

Danish atomic physicist who collaborated with Ernest Rutherford (q.v.) in Manchester in 1912 and there formulated his quantum theory of the electronic structure of the atom. He was subsequently Reader in Theoretical Physics at Manchester 1914–16.
see CAMBRIDGE, MANCHESTER

Bolsover, Thomas (*fl.* 1740s)

Invented Sheffield Plate, made by fusing a thin sheet of silver to a thicker one of copper and then rolling it out.
see SHEFFIELD

Boole, George (1815–64)

Mathematician, founder of Boolean Algebra.
see CORK

Bouch, Thomas (1822–80)

Civil engineer, builder of the original Firth of Forth Bridge (1877). In its first year it collapsed in a gale as a train was crossing: seventy-five lives were lost.
see DUNDEE

Boulton, Matthew (1728–1809)

A great industrialist, with a particular interest in steam engines. The factory he founded at Soho with James Watt (q.v.) was one of the wonders of the world. A leading spirit in the Lunar Society of Birmingham.
see BIRMINGHAM

Bowman, William(1816–92)

Pioneer of histology and a great ophthalmic surgeon. Identified Bowman's Capsule in the kidney.
see NANTWICH

Boyd Orr, John (Baron Boyd Orr 1880–1971)

A Director of the Rowett Research Institute and Professor of Agriculture, Aberdeen University. First Director of the UN Food and Agriculture Organisation (1945–8). Nobel Peace Prize 1959.
see ABERDEEN

Boyle, Robert (1727–91)

Famous as the 'Father of Chemistry and son of the Earl of Cork'. In his *Sceptical Chymist* (1662) he advanced an essentially atomistic theory of chemistry. Formulated Boyle's Law which relates the volume of a gas to its pressure.
see LISMORE, LONDON, OXFORD, STALYBRIDGE

Bradley, James (1693–1762)

Astronomer: appointed Astronomer Royal in 1742. Discovered the aberration of light, a phenomenon due to light having a finite velocity, which he calculated.
see SHERBORNE

Bradshaw, George (1801–53)

Promoter of popular railway travel. His name is

perpetuated as the original compiler and publisher of *Bradshaw's Railway Companion* (1838).
see MANCHESTER

Bragg, Lawrence (1890–1971)
Physicist, son of William Bragg (q.v.). Together, they made important contributions to the investigation of crystal structure by means of X-rays. They shared a Nobel Prize in 1915. He succeeded Ernest Rutherford (q.v.) as Cavendish Professor of Physics at Cambridge and later succeeded his father as Director of the Royal Institution in London.
see CAMBRIDGE, LONDON

Bragg, William (1862–1942)
Physicist: father of Lawrence Bragg (q.v.). After a few years as professor in Adelaide he was appointed professor at Leeds. He was subsequently Director of the Royal Institution.
see CAMBRIDGE, WIGTON

Brearley, Harry (1871–1948)
Began working life as bottle-washer in a laboratory and rose to be Technical Director of Brown Bayley Steelworks, Sheffield. Remembered for his successful promotion of stainless steel.
see SHEFFIELD

Brewster, David (1781–1868)
Physicist, remembered especially for his research on optics, particularly the polarization of light. Invented the kaleidoscope.
see EDINBURGH

Briggs, Henry (1561–1631)
Mathematician; appointed Savilian Professor of Geometry, Oxford, in 1619. Chiefly remembered for his enthusiastic promotion of John Napier's (q.v.) invention of logarithms.
see HALIFAX, LONDON, OXFORD

Bright, John (1811–89)
Radical statesman, much concerned with reform of the Corn Laws. Involved in a number of industrial enterprises, including lead mining in Wales.
see DYLIFE

Bright, Richard (1789–1858)
Physician: first to describe chronic nephritis (Bright's disease).
see BRISTOL, LONDON

Brindley, James (1716–72)
Although never more than semi-literate, Brindley had a very successful career as a canal builder, often in association with the Duke of Bridgewater.
see BARTON, BUXTON, LEEK

Brinkley, John (1763–1835)
Combined a career as astronomer with that of a priest. In 1792 he was appointed first Astronomer Royal for Ireland and in 1826 became Bishop of Cloyne.
see DUBLIN

Broom, Robert (1866–1951)
Palaeontologist, noted for his systematic unearthing of fossil remains – including those of hominids – in South Africa.
see PAISLEY

Brown, Arthur Whitten (1886–1948)
Pioneer aviator: with J.W. Alcock (q.v.) made first non-stop trans-Atlantic flight in 1919.
see MANCHESTER

Brown, Robert (1773–1858)
Botanist who worked with Joseph Banks (q.v.) and after his death in 1826 supervised the removal of Banks' huge collection to the British Museum. First person to observe the random motion of tiny particles (pollen grains) suspended in liquids (Brownian Movement).
see ABERDEEN, MONTROSE

Brown, Samuel (1776–1852)
Naval officer and civil engineer. Devised improved chain-links for ships' cables and suspension bridges.
see BERWICK-UPON-TWEED

Browne, Thomas (1605–82)
Physician, defender of right to freedom of opinion. His *Hydrotaphia*, a memoir on urn burial, is one of the treasures of the English language.
see NORWICH

Brunel, Isambard Kingdom (1806–59)
One of the greatest 19th-century engineers. He was engineer to the Great Western Railway and pioneered the building of steamships, though his *Great Eastern* was a disaster and ended up as a cable-layer.
see BRISTOL, EXETER, LONDON, MAIDENHEAD, NEWPORT, PLYMOUTH, PORTSMOUTH

Brunel, Marc Isambard (1769–1849)
Engineer: pioneer of mass-production methods in industry, notably making wooden pulley-blocks for the Royal Navy. Father of I.K. Brunel (q.v.).

Devised mechanical excavation shield for building first Thames tunnel.
see LONDON

Buckland, William (1748–1856)

Professor of Mineralogy at Oxford, noted for his personal eccentricities.
see LYME REGIS, OXFORD, RHOSSILI

Budd, William (1811–80)

Pioneer of epidemiology; demonstrated that typhoid is spread by contagion.
see BRISTOL

Budding, Edward (1741–99)

Inventor of the lawn mower.
see STROUD

Bullard, Edward (1907–80)

Pioneer of marine geophysics. Director of the National Physical Laboratory 1950–55.
see CAMBRIDGE

Burton, Decimus (1800–81)

Architect: laid out Hyde Park in London in 1825. Designed the new town of Fleetwood in 1836.
see FLEETWOOD

Burton, Robert (1577–1640)

Devoted his whole life to writing a single book, the immensely erudite *Anatomy of Melancholy* (1621) a treatise on morbid psychology. Samuel Johnson said it was the only book that ever got him out of bed two hours sooner than he wished to rise.
see LINDLEY

Cabot, Sebastian (1474–1557)

Merchant venturer, who was also a skilled cartographer and navigator.
see BRISTOL

Cadbury, George (1839–1922)

Chocolate manufacturer who introduced new technology into the industry. Among the 19th-century industrialists who integrated their factories with self-contained village communities for their workers.
see BIRMINGHAM

Caius, John (1510–73)

Physician to Edward VI, Mary and Elizabeth: nine times President of the Royal College of Physicians. First in Britain to teach practical anatomy publicly. Distinguished natural historian.
see CAMBRIDGE, NORWICH

Callan, Nicholas (d.1864)

Catholic priest who taught physics at Maynooth, with particular reference to electrical communication.
see MAYNOOTH

Cavendish, Henry (1731–1810)

Distinguished but eccentric chemist/physicist; nephew of the 3rd Duke of Devonshire. The accuracy of his determination of the density of the earth was not surpassed for a century.
see BAKEWELL, DERBY, LONDON

Cayley, George (1773–1857)

Founder of aerodynamics. He elucidated the basic principles of heavier-than-air flight but could not put them into practice as no suitable power unit was then available. The Wright brothers paid tribute to him.
see CHARD, LONDON, SCARBOROUGH

Chadwick, James (1891–1974)

Atomic physicist: noted for his research on radioactivity and discovery of the neutron.
see MACCLESFIELD, LIVERPOOL

Chain, Ernst Boris (1906–79)

German-born biochemist; chief collaborator with Howard Florey (q.v.) in research that revealed the unique therapeutic properties of penicillin. In 1945 shared Nobel Prize with Florey and Alexander Fleming (q.v.).
see OXFORD

Chambers, William (1800–83)

Printer and publisher of the great *Chambers Encyclopaedia*.
see EDINBURGH

Chester, Robert of (b. early 12th century)

A shadowy figure, but highly esteemed by his contemporaries. Introduced Arabic alchemy and mathematics to western Europe. Made first Latin translation of the Koran.
see CHESTER

Clifton, Robert (1836–1921)

Appointed Professor of Experimental Physics in Oxford – though with no great interest in experiment – in 1865. Secured funds from estate of Earl of Clarendon to build the original Clarendon Laboratory.
see OXFORD

Cobbett, William (1763–1835)

Self-taught advocate of agricultural reform: noted

political polemicist. Remembered for his *Rural Rides*.
see FARNHAM

Cobden, Richard (1804–65)
Politician and economist: apostle of free trade. Had various industrial interests, including calico printing in Manchester and lead mining in Wales.
see DYLIFE

Cocker, Edward (1631–75)
Arithmetician who taught in London. His posthumous *Arithmetick, being a Plain and Easy Method* (1678) ran through over a hundred editions. Praised in *The Apprentice* (1756) a play by Arthur Murphy.
see LONDON

Cole, Humfrey (?1530–91)
One of a group of skilled instrument makers who flourished in 16th-century London.
see LONDON

Colman, Jeremiah (1830–98)
Founder of mustard manufacture in Norwich. Claimed he made his fortune not from the mustard people ate but what they left on their plates.
see NORWICH

Constable, William (1721–91)
Botanist: elected Fellow of the Royal Society 1775.
see HULL

Conybeare, William (1787–1857)
Geologist/theologian, remembered for his masterly synopsis of British stratigraphy, published in his *Outlines of the Geology of England and Wales* (1822).
see CARDIFF

Cook, James (1728–79)
Navigator and hydrographer, famous for his exploration and charting of the Pacific Ocean. In *Endeavour* led Royal Society expedition to observe transit of Venus in Tahiti (1769).
see LONDON, WHITBY

Cooke, Thomas (1807–68)
Optician and clockmaker. Famed for his astronomical telescopes and ancillary equipment.
see YORK

Cookworthy, William (1705–80)
Pharmacist who developed porcelain manufacture in England after discovering china clay at St Austell.
see BRISTOL, KINGSBRIDGE, ST AUSTELL

Cort, Henry (1740–1800)
'The Great Finer', so-called because of his invention of the puddling process for refining wrought iron. Also introduced the use of mechanical rollers in place of traditional hammering to shape iron.
see FAREHAM, LANCASTER

Courtauld, Samuel (1798–1881)
Founder of the famous silk manufacturing company which at the beginning of this century pioneered the rayon industry in Britain.
see BRAINTREE, HALSTEAD

Crabtree, William (1610–44)
Astronomer who with Jeremiah Horrocks (q.v.) made an historic observation of the transit of Venus in 1639.
see MANCHESTER

Cragg, John (1767–1854)
Liverpool iron founder, pioneer of building construction in cast-iron.
see LIVERPOOL

Crampton, Thomas (1816–88)
Railway engineer, remembered for his long-boilered locomotives which were extremely popular on the Continent.
see BROADSTAIRS, LIVERPOOL

Crawford, William (1816–95)
Founder of palaeobotany. Remembered for his research on fossil plants of the Coal Measures.
see SCARBOROUGH

Crick, Francis (b.1916)
Molecular biologist famous for elucidating, with J.D. Watson (q.v.) the structure of DNA – the Golden Helix – which controls heredity. Nobel Laureate 1962.
see CAMBRIDGE

Crompton, Samuel (1753–1827)
Pioneer of the mechanical textile industry. In 1779 invented his famous 'mule', so called because it was a cross between Hargreaves' jenny and Arkwright's water-frame spinning machine. It was a versatile machine, suitable for spinning any type of fibre.
see BOLTON

Crookes, William (1832–1919)

Discovered the element thallium (1861) and did pioneer research on radiation and cathode rays. The Crookes radiometer is a popular scientific toy. Of him it was said '*Ubi Crookes, ibi lux*' (Where is Crookes, there is light).

see LONDON

Cullen, William (1710–90)

Practised medicine in Hamilton, then from 1747 lectured on chemistry at Glasgow – the first independent lectureship in chemistry in the British Isles. In 1755 he moved to Edinburgh, where his most famous student was Joseph Black (q.v.).

see GLASGOW, EDINBURGH

Curtis, William (1746–99)

Botanist. After setting up as an apothecary in London, he established botanic gardens successively in Bermondsey, Lambeth and Brompton. His *Flora Londoniensis* was abandoned after two huge volumes, but his *Botanical Magazine* (1787) was extremely successful. He maintained that the *Flora* earned him praise but the *Magazine* earned him pudding.

see ALTON

Dalton, John (1766–1844)

Perhaps the greatest of all chemists. Formulated theory to explain chemical reactions postulating that the atoms of different elements are distinguished by their weights. In 1803 published first table of atomic weights. On his death forty thousand people filed past his coffin.

see EAGLESFIELD, KENDAL, MANCHESTER

Dancer, John (1812–87)

Scientific instrument maker and optician; inventor of microphotography.

see MANCHESTER

Darby, Abraham (1678–1717)

Founder of a dynasty of iron founders at Telford. Revolutionized the industry by successfully smelting iron with coke rather than charcoal, supplies of which were fast diminishing.

see TELFORD

Darwin, Charles (1809–82)

Naturalist, famous for his theory of evolution by natural selection (1858), which had been independently conceived by Alfred Russel Wallace (q.v.). It aroused immense controversy and is still contentious in some states of the USA.

see CAMBRIDGE, DOWNE, LONDON, SHREWSBURY

Darwin, Erasmus (1731–1802)

Physician whose intellectual curiosity ranged over many fields: grandfather of Charles Darwin (q.v.). His best-selling *Botanic Garden* – a survey of the science of his day and an exposition of the Linnaean system of plant classification – was written in heroic couplets.

see BIRMINGHAM, LICHFIELD

Darwin, Francis (1848–1925)

Botanist. Reader in botany at Cambridge; assisted his father, Charles Darwin (q.v.), at Downe.

see CAMBRIDGE

Darwin, George (1845–1912)

Astronomer: Professor of Astronomy at Cambridge.

see CAMBRIDGE

Darwin, Horace (1851–1928)

Another gifted son of Charles Darwin (q.v.). He founded a small scientific instrument firm in 1881 which became the now well-known Cambridge Instrument Company.

see CAMBRIDGE

Daubeny, Charles (1795–1863)

Concurrently Professor of Botany, Chemistry and Rural Economy at Oxford.

see OXFORD

Davy, Humphry (1728–1829)

Talented scientist who attracted fashionable audiences to his lectures at the Royal Institution, London. Famous for his invention of the miners' safety lamp and applying the Voltaic cell to isolation of sodium and potassium.

see BRISTOL, LONDON, PENZANCE

De Ferranti, Sebastian (1864–1930)

Pioneer of the generation and distribution of high-voltage electricity for public use. Established his own factory to manufacture electrical equipment in 1896.

see LIVERPOOL

De Havilland (1882–1965)

Aircraft designer and manufacturer. Among his successes were the Tiger Moth, the Mosquito, and the Vampire jet.

see LONDON COLNEY

De La Rue, Warren (1815–89)

A member of the well-known printing firm, his real interest was in astronomy and he was among the

first to use photographic methods, especially for studying the Sun and Moon.
see GUERNSEY

De Valera, Eamon (1882–1975)
Irish statesman who had a life-long interest in mathematics, of which he was in his early years a professor. Founded the Dublin Institute of Advanced Studies (1940).
see DUBLIN, BRUREE

Dee, John (1527–1608)
Erudite and widely travelled scholar who – in the manner of his day – dabbled equally in mathematics (in which he was pre-eminent), alchemy and astrology. He was official astrologer to the court of Elizabeth.
see LONDON, MANCHESTER

Derham, William (1657–1735)
First person to attempt an accurate measurement of the speed of sound.
see UPMINSTER

Desmarest, Nicolas (1725–1815)
French geologist who made first detailed study of the Giant's Causeway, Antrim, and correctly identified it as analogous to similar formations in the Auvergne.
see BUSHMILLS

Dewar, James (1842–1923)
Physicist, noted for his research on the properties of gases at low temperatures and their liquefaction. Succeeded in liquefying hydrogen 1898. Invented the now ubiquitous vacuum flask.
see KINCARDINE-ON-FORTH

Dick, William (1793–1866)
Son of a farrier, he became a veterinarian and founded the Edinburgh Veterinary College.
see EDINBURGH

Digges, Leonard (1510–81);
Digges, Thomas (1543 –95)
Thomas Digges was a highly regarded mathematician and an authority on navigation. He was also a skilled instrument maker and, with his father Leonard, is now regarded as the true inventor of the telescope, anticipating the Dutch by more than 40 years.
see LONDON

Dirac, Paul (1902–84)
Among the most creative of all theoretical physicists. Founded quantum mechanics and formulated the relativistic wave-equation for the electron. In 1933 he shared the Nobel Prize for physics with Erwin Schrödinger (q.v.).
see BRISTOL, CAMBRIDGE

Dover, Thomas (1660–1742)
Flamboyant physician who also practised as a privateer. Remembered for 'Dover's Powder' a fearful nostrum whose composition was published in his book *The Ancient Physicians Legacy to his Country* (1733).
see BRISTOL

Doyle, Arthur Conan (1859–1930)
Best remembered as the creator of Sherlock Holmes, but also a practising physician with an interest in forensic medicine.
see EDINBURGH

Draper, John (1811–82)
Pioneer of astronomical photography. Took first photographs of the Moon in 1839/40.
see ST HELENS

Dunlap, John (1747–1812)
Printer: printed first broadside of the American Declaration of Independence.
see STRABANE

Dunlop, John Boyd (1840–1921)
Practised very successfully as a veterinary surgeon in Edinburgh and Belfast, but remembered primarily for his invention of the pneumatic tyre in 1887.
see BELFAST, COVENTRY, DUBLIN, DREGHORN

Dyson, Frank (1868–1939)
Astronomer Royal 1910–33. Organized two expeditions to observe the solar eclipse of 1919, which precisely confirmed Einstein's prediction of the deflection of light by gravity.
see ASHBY-DE-LA-ZOUCH, LONDON

Earnshaw, Thomas (1799–1829)
One of the great English watch and clockmakers. Invented the cylindrical balance spring and the detached detent escapement.
see LONDON

Eddington, Arthur (1882–1946)
Astronomer, with particular interest in stellar structure. He had a gift for popular exposition and did much to interpret contemporary science to the public.
see KENDAL

Edgeworth, Richard Lovell (1744–1817)

Educationist and versatile inventor. Popular member of the Lunar Society of Birmingham.
see BATH, BIRMINGHAM

Edwards, William (1719–89)

Civil engineer. Designed the elegant single-arch bridge over the River Taff at Pontypridd.
see PONTYPRIDD

Elkington, George (1801–65)

Pioneer of electroplating with silver, displacing Sheffield Plate. Also had copper-smelting interests in south Wales.
see BIRMINGHAM

Elsmore, Alexander (fl. 1890–1910)

Mining engineer who pioneered the important flotation process for ore concentration.
see BEDDGELERT

Fairbairn, Peter (1799–1861)

Iron founder who made important improvements in wool-processing machinery, particularly in replacing wooden parts by metal.
see LEEDS

Fairbairn, William (1789–1874)

Versatile engineer whose interests ranged from industrial machinery and water-wheels to ferry boats and bridges. He claimed that during his long career he had built nearly a thousand bridges.
see BANGOR, KELSO, MANCHESTER

Fairlie, Robert (1831–85)

Locomotive Superintendent of the Londonderry and Coleraine Railway: protagonist of narrow-gauge railways. His unorthodox double-ended locomotives aroused international interest.
see BLAENAU FFESTINIOG

Faraday, Michael (1791–1867)

One of the outstanding scientists of the 19th century, famous for his highly original research on electromagnetism and inventor of the dynamo. A great popularizer of science: his lectures at the Royal Institution attracted large fashionable audiences.
see AMLWCH, LONDON, MERTHYR TYDFIL, NEWINGTON

Ferguson, James (1710–76)

Astronomer and instrument maker. He settled in London in 1743 and gave many public lectures there and in the provinces: he became unofficial 'populariser in residence' to the court of George III.
see BANFF

Ferrier, David (1843–1928)

Neurologist: did pioneer research on the function of the brain. Appointed Professor of Neuropathology in London, 1889. Founded the journal Brain (1878). Taken to court by anti-vivisectionists but won the case for ethical animal experimentation.
see ABERDEEN

Finniston, Monty (1912–91)

Metallurgist and industrialist. At UK Atomic Energy Authority, Harwell (1948–58) was responsible for producing Britain's first plutonium. Chairman of British Steel Corporation 1973–6.
see GLASGOW

Fitzgerald, George Francis (1851–1901)

Theoretical physicist: advanced the Fitzgerald Contraction theory to explain why the measured speed of light is not affected by the velocity of its source.
see DUBLIN

Flamsteed, John (1646–1719)

First Astronomer Royal (1675). Working single-handed, with his own instruments, he compiled a star catalogue recording over twenty thousand observations. Fell into dispute with Newton (q.v.) by refusing to provide data Newton needed to complete his lunar theory.
see BURSLOW, DERBY

Fleetwood, Peter Hesketh (1801–66)

Originally Peter Hesketh but changed his name to Fleetwood on building the town of that name in 1836.
see FLEETWOOD

Fleming, Alexander (1881–1955)

Bacteriologist, remembered for two major medical discoveries – lysozyme (1922) and penicillin (1928). For various reasons, he failed to recognize the unique therapeutic properties of penicillin and the development of this fell to Howard Florey and Ernst Chain (qq.v.) at Oxford. Nobel Laureate 1945.
see LONDON

Florey, Howard (Baron Florey 1898–1968)

Australian pathologist. Discoverer, with his colleagues at Oxford, of the remarkable therapeutic properties of penicillin and initiated its large-scale production. Nobel Laureate 1945. Fiftieth President of

the Royal Society 1960–5.
see LONDON, OXFORD, SHEFFIELD

Fludd, Robert (1574–1637)

Physician/alchemist. He spurned the teaching of traditional classical writers such as Aristotle and Galen. Although not averse to experimental evidence he sought to interpret the Universe on the evidence of the Bible.
see BEARSTEAD

Forsyth, Alexander (1779–1843)

Chemist. Inventor of the percussion cap (mercury fulminate) for firearms.
see ABERDEEN

Foulis, Robert (1707–76)

Printer to Glasgow University (1743) notable for the very high quality of his work.
see GLASGOW

Fowler, John (1817–98)

Civil engineer. Builder of the Firth of Forth railway bridge: did much work on the London underground railway.
see BRIDGNORTH, SHEFFIELD, SOUTH QUEENSFERRY

Frankland, Edward (1825–99)

Chemist – latterly at the Royal School of Mines, London – remembered for research on the combining powers of the elements (valency). Interested also in practical problems, he was a consultant on coal gas and water supply.
see MANCHESTER

Franklin, Benjamin (1706–90)

American statesman: one of the five men who drafted the Declaration of Independence (1776). No less famous for his electrical experiments, relating particularly to atmospheric electricity. It was said that 'he found electricity a curiosity and left it a science'.
see LONDON

Friese-Green, William (1855–1921)

Pioneer of cinematography.
see BRISTOL, LONDON

Fry, John (1728–87)

Leading member of the Fry family, who established chocolate manufacture as a major industry.
see BRISTOL

Galton, Francis (1822–1911)

Scientist of wide interests, including meteorology, but in later life particularly interested in genetics in relation to the scientific breeding of human populations. Founder of eugenics.
see BIRMINGHAM, LONDON

Geikie, Archibald (1835–1924)

Geologist, in his day the leading figure in his field. A man of wide learning, he shared with C.N. Hinshelwood (q.v.) the distinction of being simultaneously President of both the Royal Society and the Classical Association.
see OXFORD

Gellibrand, Henry (1593–1637)

Professor of Astronomy in Gresham College, London. Did much to bring Napier's logarithms to public notice.
see LONDON

Gemini, Thomas (*fl.* 1524–62)

Skilled instrument maker. Engraved magnificent plates for the English edition (1545) of Vesalius' *Anatomy*.
see LONDON

Gerard, John (1545–1612)

Herbalist, author of the famous *Herball* of 1597, illustrated by eighteen hundred woodcuts of native and imported plants. It is a valuable survey of botanical knowledge and outlook at the end of the 16th century.
see NANTWICH

Gilbert, Joseph Henry (1817–1901)

Agricultural chemist: Director (with J.B. Lawes (q.v.)) of the Rothamsted Laboratory. Professor of Rural Economy, Oxford, 1884–90.
see ROTHAMSTED

Gilbert, William (1544–1603)

Successful medical practitioner in London, but mostly remembered for his experimental study of electricity and magnetism. Refuted alchemical belief in transmutation of metals.
see CAMBRIDGE, COLCHESTER, LONDON

Gilchrist, Percy Carlyle (1851–1935)

Inventor, with his cousin S.G. Thomas (q.v.), of the important basic process for dephosphorizing steel, thus opening to exploitation the vast ore deposits of Lorraine and Luxemburg.
see LYME REGIS

Gill, David (1843–1914)

For ten years a working watchmaker, he developed

a talent for astronomy, becoming Astronomer of the Cape of Good Hope.
see ABERDEEN

Gillott, Joseph (1799–1873)
Industrialist who made a large fortune by developing machinery to manufacture steel pen-nibs.
see BIRMINGHAM

Glaisher, James (1809–1903)
Mathematician, with particular interest in the history of mathematics. Editor of *Quarterly Journal of Mathematics*.
see LONDON

Goodricke, John (1764–86)
Infant prodigy who despite being stricken deaf and dumb, and dying at 22, made notable contributions to astronomy.
see YORK

Gould, John (1804–81)
Ornithologist and publisher, remembered for his magnificent bird illustrations.
see LYME REGIS

Graham, Thomas (1805–69)
Chemist, with particular interest in the physical properties of gases, including their diffusion and solubility. Also made an intensive study of colloids. First President of the Chemical Society (1841).
see GLASGOW, LONDON

Graves, Robert (1796–1853)
Physician and acute clinical observer. He described exophthalmic goitre in 1835, a symptom of thyrotoxicosis (Graves' disease). He asked that his epitaph be 'he fed fevers'.
see DUBLIN

Gray, Henry (1827–61)
Surgeon and anatomist. His *Anatomy, Descriptive and Surgical* (1858) went through many editions and was the standard British textbook for more than a century.
see LONDON

Greatorex, Ralph (1625–1712)
A leading London maker of scientific instruments.
see LONDON

Green, George (1793–1841)
Self-taught mathematician remembered for his theoretical work on electricity and magnetism. He

also studied wave motion and the propagation of light.
see CAMBRIDGE

Greg, Samuel (1758–1834)
Built Quarry Bank Mill, Styal (1784), as both a textile mill and a communal village for his employees.
see STYAL

Gregor, William (1761–1817)
Chemist: first to recognize titanium as a new element.
see CREED, ST AUSTELL

Gregory, James (1638–75); Gregory, David (1659–1708); Gregorie, John (1724–73)
Members of an Aberdeen family who made notable contributions in the fields of astronomy, mathematics and mechanics.
see ABERDEEN

Gresham, Thomas (1519–79)
Diplomat and financier. In London built the Royal Exchange and Gresham College, first home of the Royal Society.
see LONDON

Gresley, Nigel (1876–1941)
Steam-locomotive engineer. Among those he designed was *Mallard*, which in 1938 achieved the world record speed for a steam locomotive of 126 mph.
see EDINBURGH

Grossetest, Robert (1175–1253)
One of the greatest medieval scholars, who took all knowledge as his province. He wrote profusely on subjects ranging from the calendar to comets, optics to heat, sound to astronomy.
see LINCOLN, OXFORD

Gunter, Edmund (1581–1626)
Professor of Astronomy in Gresham College. Publicized Napier's logarithms.
see LONDON

Hackworth, Timothy (1786–1850)
Manager of the Stockton and Darlington Railway 1825–40. Built several famous locomotives, including *Puffing Billy* and *Sans Pareil*.
see SHILDON

Hadley, John (1682–1744)
Maker of mathematical instruments, including the

reflecting quadrant (ancestor of the modern sextant).
see BARNET

Hakluyt, Richard (1522–1616)
Geographer. Wrote widely on navigation and exploration, notably his *Principal Navigations, Voyages, and Discoveries of the English Nation* (1598–1600).
see BRISTOL

Hales, Stephen (1677–1761)
Chemist and plant physiologist: one of the first to recognize the importance of weighing and measuring in experiments. Investigated the role of air, water and light, in the life of plants and the circulation of the blood in animals.
see BEKESBOURNE, LONDON

Halley, Edmond (1656–1742)
Astronomer: first to compile a systematic catalogue of the stars of the Southern Hemisphere. Studied the periodicity of comets: remembered for Halley's Comet, with a periodicity of 76 years. Investigated the earth's magnetism.
see OXFORD

Hamilton, William Rowan (1805–65)
Mathematician: an infant prodigy with remarkable capacity for involved mental calculations. Predicted existence of conical refraction of light. Invented quaternions.
see DUBLIN

Hancock, Thomas (1786–1865)
A founder of the modern rubber industry. Invented a vulcanization process for hardening rubber by treatment with sulphur.
see MARLBOROUGH

Hargreaves, James (1720/1–78)
Textile engineer: inventor of the spinning jenny. This could spin eight threads simultaneously and led to riots by the threatened handloom weavers.
see BLACKBURN, NOTTINGHAM

Harland, Edward (1831–95)
A founder of the famous shipbuilding company Harland and Wolff.
see BELFAST

Har(r)iot, Thomas (1560–1621)
Astronomer, physicist, algebrist. Discovered sunspots independently of Galileo and probably the law of refraction before Willebrord Snell. Improved algebraic notation.
see LONDON

Harris, Joseph (1602–64)
Astronomer: wrote major works on navigation and navigational instruments.
see TALGARTH

Harrison, John (1693–1776)
Horologist: constructed the first chronometer capable of being used at sea to determine longitude. For this he received awards totalling £20,000.
see FOULBY, LONDON

Harvey, William (1578–1657)
Studied medicine in Padua and established himself as a fashionable physician in London. Discovered the circulation of the blood (published 1628) initially arousing much controversy. Also made important contributions to embryology, using incubating hens' eggs as his experimental material.
see CAMBRIDGE, CANTERBURY, FOLKESTONE

Hauksbee, Francis (1666–1713)
Student of Robert Boyle (q.v.) for whom he designed and constructed air-pumps. Showed that air at low pressure glows when subjected to an electric discharge.
see COLCHESTER

Hawksley, Thomas (1807–93)
Water engineer responsible for major schemes in Liverpool, Nottingham, Sheffield, Leicester and elsewhere.
see LONDON

Haworth, Adrian (1768–1833)
For many years his *Lepidoptera Britannica* was the standard work on British butterflies.
see HULL

Haworth, Walter (1883–1950)
Pioneer of carbohydrate chemistry. Interest in sugars led him to investigate vitamin C: with colleagues at Birmingham he elucidated its structure and subsequently synthesized it. Nobel Laureate 1947.
see BIRMINGHAM

Heathcote, John (1783–1861)
Inventor of the bobbin-net machine for lace-making. Luddites destroyed his factory in Loughborough in 1816 and he moved to Devon.

Also invented a silk-cocoon reeling machine and a steam plough.
see TIVERTON

Heaviside, Oliver (1850–1925)

By his theoretical approach he provided the basis for considerable improvements in telegraphic equipment. Discovered a layer in the upper atmosphere (Heaviside layer) which reflects radio waves.
see PAIGNTON

Henry, Thomas(1734–1816)

Practised in Manchester as apothecary and surgeon. Lectured on chemistry, bleaching and dyeing to Manchester College of Arts and Science.
see MANCHESTER

Henslow, John (1796–1861)

Professor of Botany, Cambridge 1827–61. Powerfully influenced the career of Charles Darwin (q.v.) by recommending him as naturalist for the *Beagle* expedition. Also active field geologist.
see CAMBRIDGE

Henson, William (1805–88)

Lace manufacturer who, with John Stringfellow (q.v.), developed an active interest in heavier-than-air flight. Together they designed an ambitious, but unsuccessful, aerial steam carriage.
see CHARD

Herschel, John (1792–1871)

Distinguished astronomer, physicist and mathematician. Son of William Herschel (q.v.) whom he assisted in his observations. Spent four years at the Cape compiling catalogue of the stars of the Southern Hemisphere. Master of the Mint, 1830.
see LONDON

Herschel, William (1738–1822)

Astronomer who began career as fashionable organist in Bath: from 1766 turned to astronomy and built his own telescope. Discovered planet Uranus, the shape of the Milky Way, and the intrinsic motion of the Sun through space.
see BATH, LONDON

Hewish, Antony (b.1924)

Radio astronomer, first to identify the 'twinkling' stars known as pulsars (1967). Nobel Laureate 1974.
see CAMBRIDGE, ST AUSTELL

Higgins, William (1763–1825)

Chemist to the Irish Linen Board 1795–1822. Vociferously claimed to have anticipated the atomic chemical theory of John Dalton (q.v.) but this is not now accepted.
see DUBLIN

Hill, Archibald Vivian (1886–1977)

Physiologist, particularly interested in energy exchanges in nerves and muscles. As President of the British Association for the Advancement of Science (1952) was among first to draw attention to the ethical and scientific problems of over-population.
see BRISTOL

Hinshelwood, Cyril Norman (1897–1967)

Physical chemist with a particular interest in the kinetics of chemical reactions. He was a man of many talents. Except for Archibald Geikie (q.v.) he was the only person to be simultaneously President of the Royal Society and the Classical Association. An authority on Chinese ceramics and oriental rugs, and a gifted linguist. Nobel Laureate 1956.
see OXFORD

Hobbes, Thomas (1588–1679)

Though primarily remembered as a political philosopher, whose masterpiece was *Leviathan* (1650), Hobbes was also an accomplished mathematician and wrote on optics, holding a corpuscular theory of the nature of light. Career much influenced by long association with the Cavendish family.
see AULT HUCKNALL, MALMESBURY

Hodgkin, Alan (b.1914)

Physiologist, with a particular interest in the mechanism of the transmission of impulses along nerve fibres. Nobel Laureate 1963.
see BANBURY, CAMBRIDGE

Hodgkin, Dorothy (1910–94)

Chemist: pioneer of X-ray crystallographic analysis. Two of her most spectacular successes were with penicillin and vitamin B-12. Only the third woman to be awarded a Nobel Prize (1964).
see OXFORD

Holloway, Thomas (1800–83)

Philanthropist who made his fortune from patent medicines. Founded Royal Holloway College, 1876.
see EGHAM

Holmes, Arthur (1890–1965)
Geologist, with a particular interest in igneous rocks and their origin. Used radioactivity measurements to estimate the age of the earth.
see HEBBURN

Hooke, Robert (1635–1702)
A brilliant and versatile scientist, his research included astronomy, optics, geology, microscopy and the elastic properties of materials. Appointed a secretary of the Royal Society in 1677.
see FRESHWATER, OXFORD

Hooker, Joseph Dalton (1817–1911)
Plant taxonomist: appointed Director of Kew Gardens 1865 in succession to his father W.J. Hooker (q.v.).
see HALESWORTH

Hooker, William Jackson (1785–1865)
His discovery of a rare moss in 1805 led him to a career in botany, ultimately as first Director of Kew Gardens (1841).
see GLASGOW

Hopkins, William Gowland (1861–1947)
Biochemist, remembered for research on proteins and the chemistry of muscular contraction. Study of the relationship between growth and diet led him to the concept of 'accessory food factors' (vitamins).
see CAMBRIDGE, EASTBOURNE

Hornby, Frank (1863–1937)
Inventor of Meccano, the model engineering toy which started many engineers on their careers.
see LIVERPOOL

Horrocks, Jeremiah (1618–41)
Made astronomical history in 1639 by predicting and observing a transit of Venus which had been overlooked by others because of an error in the Rudolfine Tables.
see MUCH HOOLE

Hounsfield, Godfrey (b.1919)
Inventor of computer-assisted tomography (body scanner). Nobel Laureate 1981.
see NEWARK

Hunter, John (1728–93);
Hunter, William (1718–83)
Scottish physicians. John Hunter began life as a cabinet-maker and rose to be a very successful surgeon: he has been called 'the Father of Scientific Surgery'. His brother William, with whom he

originally collaborated but later quarrelled, had an equally successful career as anatomist and obstetrician: he founded a private medical school in Great Windmill Street, London, in 1767.
see GLASGOW, LONDON

Hurter, Ferdinand (1844–98)
Chemist who contributed much to the development of the chemical industry on Merseyside.
see WIDNES

Hutton, James (1726–97)
Founder of geology as a science in its own right. His *Theory of the Earth* (1795) is one of the classics of scientific literature.
see COCKSBURNSPATH

Huxley, Andrew (b.1917)
Physiologist: a member of the scientifically gifted Huxley family. Investigated nervous transmission and outlined a theory of muscular contraction. Nobel Laureate 1963.
see CAMBRIDGE

Huxley, Thomas (1825–95)
Often remembered only as 'Darwin's Bulldog' for fiercely championing his theory of evolution, but nevertheless a distinguished scientist in his own right. His research, much of it taxonomic, extended over many fields of botany and zoology.
see LONDON, OXFORD

Isaacs, Alick (1921–67)
Virologist, discoverer of interferon.
see GLASGOW, LONDON

Jenner, Edward (1749–1823)
Practised medicine in Gloucestershire after studying under John Hunter (q.v.). Pioneer of vaccination against smallpox, for which he eventually received public awards totalling £30,000.
see BERKELEY

Joly, John (1857–1933)
Geologist and physicist. Devised two methods for estimating the age of the earth, one based on the salinity of the oceans, the other on the radioactivity of rocks.
see BELFAST, DUBLIN

Jordan, Joseph (1787–1823)
Physician: founded the first provincial school of anatomy (1814).
see MANCHESTER

Josephson, Brian (b.1940)

Theoretical physicist: invented tunnelling
superconductors. Nobel Laureate 1973.
see CARDIFF

Joule, James Prescott (1818–89)

Physicist, remembered for precise measurement of
the mechanical equivalent of heat. From 1847 he
collaborated with William Thomson (Lord Kelvin)
(q.v.) in research on thermodynamics.
see MANCHESTER

Kane, Robert (1804–90)

Chemist and educationist: his *Elements of
Chemistry* was the leading textbook of the day.
Publication of his *Industrial Resources of Ireland*
led to his becoming adviser to the Irish government
on matters relating to industry and education.
see DUBLIN

Kay, John (1704–64)

Inventor of textile machinery, notably the flying
shuttle. He was ruined by litigation protecting his
patent against infringement, his house was wrecked
by Luddites, and he died a pauper in France.
see BURY

Keir, James (1735–1820)

Founder of the scientific chemical industry in
Britain.
see WEST BROMWICH

Keith, Arthur (1866–1955)

Physical anthropologist, known for his research on
fossilized hominids.
see DOWNE

Kendrew, John (b.1917)

Molecular biologist. With Max Perutz (q.v.) he
elucidated the structure of myoglobin, an important
muscle component, in 1960, using X-ray diffraction
techniques. First Director of European Molecular
Biology Laboratory, Heidelberg, 1975–82. Shared
Nobel Prize with Perutz in 1962.
see CAMBRIDGE

Kipping, Frederic (1863–1959)

Pioneer of the organic chemistry of silicon, paving
way to modern silicone plastics, remarkable for their
water-resistance and high-temperature stability.
see NOTTINGHAM

Kirwan, Richard (1733–1812)

Chemist and mineralogist. Initially a convinced
believer in phlogiston as the active principle in

combustion, he was eventually converted to
Lavoisier's view of the role of oxygen. A notable
eccentric, he wore top-coat and hat indoors, and
lived on ham and milk.
see DUBLIN

Kratzer, Nicholas (1486–1550)

One of the leading scientific instrument makers in
London. Some of his instruments are depicted in
Holbein's painting *The Ambassadors*. Official
astronomer and horologist to Henry VIII.
see LONDON

Krebs, Hans (1900–81)

Biochemist: remembered for his elucidation
of principal metabolic pathways in the cell, notably
those concerned with utilization of energy (Krebs
Cycle). Nobel Laureate 1953.
see OXFORD, SHEFFIELD

Lanchester, Frederick (1868–1946)

One of the first to apply scientific principles to
motor-car design and construction.
see BIRMINGHAM

Larmor, Joseph (1857–1942)

Lucasian Professor of Mathematics, Cambridge
1903–32. Worked on mathematical problems in
electrodynamics and thermodynamics.
see MAGHERAGALL

Lawes, John Bennet (1814–99)

Agricultural scientist, founder of the famous
research centre at Rothamsted, part of his family
estate. Pioneered use of synthetic fertilisers and
established his own factory for their manufacture.
see ROTHAMSTED

Lever, William (Lord Leverhulme 1851–1925)

Soap-maker and philanthropist. Introduced use of
vegetable oils in place of the traditional tallow.
see PORT SUNLIGHT

Lewis, Timothy (1841–86)

After qualifying in London, joined the Army
Medical Service and went to India. Made major
contributions in tropical medicine, especially in the
realm of blood parasites.
see CARMARTHEN

Lhuyd, Edward (1660–1709)

Professor of Chemistry at Oxford and assistant in
the Ashmolean Museum.
see CARDIGAN

Linacre, Thomas (1460–1524)
Physician: founder of the Royal College of
Physicians.
see CAMBRIDGE

Lindemann, Frederick (Lord Cherwell 1886–1957)
Physicist. During First World War studied problems
of flight at Royal Aircraft Establishment and
personally proved correct his theory of how to
extricate an aeroplane from spin. From 1933
personal scientific adviser to Winston Churchill.
see OXFORD

Lister, Joseph (Lord Lister 1827–1912)
Founder (1865) of antiseptic surgery, which
greatly reduced post-operative mortality. His
technique was quickly adopted world-wide.
see EDINBURGH, GLASGOW, LONDON

Lloyd, Humphrey (1800–81)
Physicist, remembered for research on optics and
magnetism.
see DUBLIN

Lockyer, Joseph (1836–1920)
Pioneer of astrophysics, who made a special study
of sunspots. Ahead of his time, he suggested that in
the intense heat of the Sun atoms may disintegrate.
see RUGBY, SIDMOUTH

Lodge, Oliver (1851–1940)
Made important contributions to theory of
propagation of electromagnetic waves, with
practical applications in telegraphy. In later life
became interested in psychic phenomena.
see BIRMINGHAM, PENKHULL

Lombe, Thomas (1685–1739)
Silk throwster, who improved machinery for
spinning silk.
see DERBY

Lonsdale, Kathleen (1903–71)
Pioneer of X-ray crystallography: Professor of
Chemistry, University College, London, 1946–68.
In 1945 was one of the first two women to be
elected to Fellowship of the Royal Society.
see NEWBRIDGE

Lovell, Bernard (b.1913)
Professor of Radio Astronomy, Manchester
University, 1951–80. His visible memorial is the
great radio telescope at Jodrell Bank.
see JODRELL BANK, MANCHESTER

Lower, Richard (1631–91)
Physician and anatomist: pioneer of blood
transfusion. In 1667 performed a successful
animal-man transfusion. His *Tractatus de corde*
(Treatise on the heart) (1669) is a classic.
see BODMIN

Lyell, Charles (1797–1875)
Established and interpreted the principles of
geology. His fame rests largely on his masterly
Principles of Geology (1830–33) which appeared in
twelve editions.
see KIRRIEMUIR

McAdam, John Loudon (1756–1836)
Road engineer. His memorial is the macadamized
(sic) system of road construction widely adopted in
Europe by the end of the 19th century.
see AYR, BRISTOL

MacEwen, William (1848–1924)
Pioneer, with Joseph Lister (q. v) of the technique
of antiseptic surgery.
see ROTHESAY

MacIntosh, Charles (1766–1843)
Chemist: inventor and manufacturer of rubberized
fabric. Mackintosh (sic) became a generic name for
waterproof coats of all kinds.
see GLASGOW

MacLaurin, Colin (1698–1746)
Mathematician and actuary: did original work of
his own in developing Newton's work on geometry,
fluxions and gravitation.
see KILMODAN

MacLeod, John James Rickard (1876–1935)
Physiologist, remembered for his pioneer work,
with F.G. Banting (with whom he shared a Nobel
Prize in 1927), on the isolation of insulin (1921).
see ABERDEEN, CLUNY

McNaught, William (1813–81)
Steam engineer. As James Watt (q.v.) had greatly
improved the Newcomen engine, so McNaught
improved Watt's by introducing a high-pressure
cylinder half-way along the beam. Many Watt
engines were 'compounded' in this way and it
became standard for most new ones.
see PAISLEY

MacNeven, William (1763–1841)
Physician. After qualifying in Vienna he

participated in the Irish Revolution (1797–8), served with the Irish Brigade of the French Army, and finally settled in New York in 1808 as Professor of Chemistry and Obstetrics, College of Physicians and Surgeons.
see BALLYNAHOWNE

Madocks, William (1773–1828)

Welsh industrialist who built the port of Portmadoc, primarily to serve the burgeoning Welsh slate industry.
see TREMADOC

Malthus, Thomas (1766–1834)

Pioneer of population science and economics, expounded in his *Essay on the Principle of Population* (1798). Charles Darwin (q.v.) was much impressed by his idea of conflict between rate of population increase and quantity of food supply.
see BATH, DORKING

Manson, Patrick (1844–1922)

Physician who spent much of his working life in the Far East, where he did important research on insect-borne diseases, notably malaria. Founded London School of Hygiene and Tropical Medicine.
see ABERDEEN, LONDON

Mantell, Gideon (1790–1852)

Surgeon/geologist. Discovered first fossil dinosaur.
see LEWES

Marconi, Guglielmo (1874–1937)

Italian pioneer of wireless communication and founder of the international Wireless Telegraph Company (1897).
see CARDIFF, LIZARD, LONDON

Martin, Archer John Porter (b.1910)

Chemist: co-inventor of the powerful analytical technique known as partition chromatography, later developed as paper chromatography. For this he and his colleague R.L.M. Synge (q.v.) were awarded a Nobel Prize in 1952.
see LEEDS

Mason, Josiah (1795–1881)

Philanthropist who made his huge fortune by manufacturing steel pen-nibs. Associated with George Elkington (q.v.) in electroplating and other ventures. Founded Mason College, now Birmingham University.
see BIRMINGHAM

Maxim, Hiram (1840–1916)

Remembered particularly for his machine-gun, but a compulsive inventor in many other fields.
see HARTLEBURY, LONDON

Maxwell, James Clerk (1831–79)

Mathematical physicist: first Cavendish Professor at Cambridge, and one of the greatest scientists of the 19th century. His great *Treatise on Electricity and Magnetism* appeared in 1873.
see ABERDEEN, CAMBRIDGE, EDINBURGH, LONDON, NEW GALLOWAY

Mayow, John (1641–1679)

Chemist/physiologist. Correctly perceived, though on philosophical rather than strictly experimental grounds, that air contains some principle – identified a century later as oxygen – that sustains both respiration and combustion.
see MARVAL

Mercer, John (1791–1866)

Pioneer of textile chemistry. Invented process for giving lustre to cotton by treatment with caustic soda (mercerizing).
see BRADFORD, GREAT HARWOOD

Michell, John (?1724–93)

Astronomer: estimated distances of the stars and discovered double stars.
see THORNHILL

Miller, Hugh (1802–56)

Apprenticeship to a stone mason aroused his interest in fossils and he embarked on a successful, though controversial, career in geology, which he described as 'the most poetical of all the sciences'.
see CROMARTY

Morris, William (Viscount Nuffield 1877–1963)

A founder of the British motor-car industry. A generous philanthropist, especially in respect of Oxford University.
see NUFFIELD, OXFORD

Moseley, Henry (1887–1915)

Pioneer atomic physicist, whose promising career came to an untimely end at Gallipoli. First to realize that the chemical properties of an atom depend on its nuclear charge rather than its weight.
see CAMBRIDGE, OXFORD

Mott, Nevill (b.1905)

Theoretical physicist, noted particularly for

research on the properties of metals, applying quantum theory to the free electrons in their structures. Nobel Laureate 1977.
see BRISTOL, CAMBRIDGE

Mudge, Thomas (1717–94)
One of the leading horologists of his day. Invented the lever escapement, much used in watches.
see EXETER

Murdock, William (1754–1839)
Pioneer of the coal-gas industry, particularly in association with the firm of Boulton and Watt in Birmingham.
see BIRMINGHAM, CAMBORNE, AUCHINLECK

Murray, Matthew (1765–1826)
Engineer: effected many improvements in flax-spinning machinery, built railway locomotives and installed power in a steam-boat.
see LEEDS

Mushet, David (1772–1847); Mushet, Robert (1811–91)
Ironmasters in the Forest of Dean, remembered particularly for their development of tough alloy steels for the machine-tool industry.
see FOREST OF DEAN

Muspratt, James (1793–1886)
Chemical manufacturer, much involved in the literary, dramatic and social life of Dublin. In 1823 moved to Liverpool and began to manufacture soda: regarded as 'the father of the British alkali industry'.
see DUBLIN

Napier, John (1550–1617)
A versatile inventor of mechanical devices for use in war but remembered particularly for his invention of logarithms (1615) which immensely simplified mathematical computations. Also introduced numbered rods ('Napier's bones') for mechanically performing multiplication and division (1617).
see EDINBURGH

Nasmyth, James (1808–90)
Manufacturer of machine tools and steam locomotives. Remembered particularly for his steam hammer (1839) so powerful, yet so delicate, that it could equally forge a massive iron billet or crack an egg.
see ECCLES, PENSHURST

Newcomen, Thomas (1663–1729)
Mining engineer, inventor of the first practical steam engine, used initially for mine drainage and pumping water but later the mainstay of the Industrial Revolution.
see DARTMOUTH, LONDON

Newton, Isaac (1642–1729)
Arguably the greatest scientist of all time, the dominant influence in scientific thought for nearly two centuries. His masterpiece was the *Principia* of 1687, incorporating his three laws of motion. He demonstrated his mastery of experiment in discovering that white light is a mixture of seven differently coloured ones: his *Optics* appeared, in English, in 1704. President of the Royal Society 1703–27. Throughout his life engaged in acrimonious disputes with his contemporaries.
see CAMBRIDGE, COLSTERWORTH, GRANTHAM, LONDON

Norton, Thomas (d. c. 1480)
Alchemist, whose *Ordinall of Alchimy* (1477) was reprinted by Elias Ashmole (q.v.) as late as 1652.
see BRISTOL

Ogg, William (1891–1979)
Agricultural scientist: Director of Rothamsted Agricultural Station 1943–58.
see ABERDEEN

Oliver, William (1695–1764)
Fashionable physician, remembered for Bath Oliver biscuits.
see BATH

Otley, Jonathan (1766–1856)
Geologist and instrument maker.
see KESWICK

Oughtred, William (1575–1660)
Mathematician and instrument maker. Wrote *Clavis Mathematica* (1643). As Vicar of Albury he accepted a succession of young gentlemen as students.
see ALBURY

Owen, Robert (1771–1858)
Textile manufacturer and railway speculator. Founded Owens College, now Manchester University.
see MANCHESTER

Parkes, Alexander (1813–90)
Versatile metallurgist: a founder of the plastics industry.
see BIRMINGHAM

Parr, Benjamin (*fl.* 1830s)
Inventor of textile machinery to recycle woollen textiles by conversion to shoddy for respinning.
see BATLEY

Parsons, Charles (1854–1931)
Son of William Parsons, 3rd Earl of Rosse (q.v.). Mechanical engineer who invented and developed the steam turbine for ship propulsion and driving high-speed dynamos.
see BIRR, NEWCASTLE UPON TYNE, LONDON

Parsons, William (3rd Earl of Rosse 1800–67)
Astronomer: constructed and operated his own telescopes, including a 36-inch reflector, at Birr Castle. Observed galaxies and binary and secondary stars.
see BIRR

Partington, James Riddick (1886–1965)
Historian of science. His four-volume *History of Chemistry* is the standard work throughout the world.
see BOLTON

Paxton, Joseph (1803–68)
Gardener and architect. Designed and built the prefabricated Crystal Palace for the Great Exhibition of 1851.
see BAKEWELL, LONDON

Pepys, Samuel (1633–1703)
Best remembered for his *Diary* but also active in the scientific world of his day: Fellow of the Royal Society 1664.
see LONDON

Perkin, William Henry (1838–1907)
Chemist. While studying at the Royal College of Chemistry, London, set up his own laboratory at home and there – at the age of 19 – prepared mauvein, the first synthetic dye. Founder of the modern dyestuffs industry.
see OXFORD

Perkin, William Henry Jr. (1860–1929)
Chemist: son of William Henry Perkin (q.v.). Had distinguished academic career as Professor of Chemistry at Manchester and Oxford. His research was particularly concerned with natural products.
see MANCHESTER, OXFORD

Perutz, Max (b.1914)
Molecular biologist; used X-ray crystallography to elucidate the structure of the blood pigment haemoglobin. Shared Nobel Prize with John Kendrew (q.v.) in 1962.
see CAMBRIDGE

Phillips, John (1800–74)
One of the founders of the British Association for the Advancement of Science (York, 1931). First Keeper of the University Museum, Oxford.
see OXFORD

Phillips, Peregrine (b.*c.* 1800)
Invented and patented the extremely important contact process for the manufacture of sulphuric acid, though it was never worked (and then only after independent invention) until the end of the century.
see BRISTOL

Pilkington, William (1800–72)
An early partner in the St Helens Crown Glass Company, now Pilkington Brothers.
see ST HELENS

Pitt-Rivers, Augustus (1827–1900)
Soldier and archaeologist, whose huge personal collection is the basis of the Pitt-Rivers Museum, Oxford.
see OXFORD

Plimsoll, Samuel (1824–98)
Social reformer: 'The Sailors' Friend'. Responsible for introduction of the Plimsoll line to ensure the safe loading of ships.
see BRISTOL, LONDON, SHEFFIELD

Porter, George (Lord Porter, b.1920)
Physical chemist. Noted for his research on the kinetics of very fast reactions. Nobel Laureate 1967.
see CAMBRIDGE

Porter, Rodney (1917–85)
Chemist, remembered for his research on protein chemistry and immunology. Nobel Laureate 1972.
see OXFORD

Powell, Cecil Frank (1908–69)
Atomic physicist particularly noted for his research on cosmic rays, using balloons to explore the upper atmosphere. Nobel Laureate 1950.
see BRISTOL

Priestley, Joseph (1733–1804)
Chemist, educationist, theologian and champion of freedom. Remembered for his research on many different gases; in 1774 prepared oxygen by heating mercuric oxide. Of him Humphry Davy (q.v.) said:

'No single person ever discovered so many new and curious substances.' Baron Cuvier said: 'He was the father of modern chemistry but would never acknowledge his child.'

see BIRMINGHAM, CALNE, LEEDS, LONDON, MANCHESTER

Prout, William (1785–1850)

Chemist: formulated Prout's Hypothesis, that all atomic weights are whole numbers. Precise contemporary analysis proved him wrong, but the discovery of isotopes in the 20th century vindicated him in a way.

see BRISTOL

Radcliffe, John (1652–1714)

Successful London physician, despite being notoriously outspoken. Endowed the Radcliffe Infirmary and Observatory in Oxford and bequeathed funds to enlarge St Bartholomew's Hospital, London.

see OXFORD

Ramsay, William (1852–1916)

Chemist, famous for his discovery of a previously unsuspected family of elements, the rare gases of the atmosphere. Nobel Prize for chemistry 1904.

see BRISTOL, LONDON

Ramsden, Jesse (1735–1800)

Engraver and skilled maker of scientific instruments, especially those of an optical nature.

see HALIFAX

Rankine, William (1820–72)

Engineer who made major contributions in many different fields, including metal fatigue, refrigeration, water supply and the stability of ships. Also did theoretical work in thermodynamics.

see GLASGOW

Ransome, Robert (1795–1864)

Well-known member of agricultural machinery manufacturing firm in Ipswich. Patented various improvements to design of ploughshares.

see IPSWICH

Ray, John (1628–1705)

Systematist of botany and zoology: variously described as 'The Father of Natural History' and 'The Aristotle of England'. His major work was the three-volume *Historia Plantarum* (1686–1704).

see BRAINTREE

Rayleigh, Lord (John William Strutt 1842–1919)

Mathematical physicist. Succeeded James Clerk Maxwell (q.v.) as Cavendish Professor, Cambridge in 1879. Collaborated with William Ramsay (q.v.) in research on the rare gases of the atmosphere. Awarded Nobel Prize for physics 1904.

see CAMBRIDGE, CHELMSFORD

Recorde, Robert (c. 1510–58)

Mathematician. Being written in English, his mathematical textbooks (1542–57) – in edition after edition – suited the new generation of mathematicians. Died in King's Bench Prison.

see TENBY

Rennie, John (1761–1821)

One of the great civil engineers of the Industrial Revolution. Architect of three famous London bridges – Southwark, Waterloo and the new London. Also engaged on many major harbour and dock schemes in Britain and abroad.

see BATH, HADDINGTON, LONDON

Reynolds, Osborne (1842–1912)

Physicist, famous for his theoretical and practical research on hydrodynamics. Reynolds' Number is a basic factor in all studies of viscous flow.

see BELFAST

Ricardo, David (1772–1823)

Economist, whose scientific approach to economic problems made him a fortune on the London Stock Exchange. Remembered for his *Principles of Political Economy and Taxation* (1817).

see HARDENHUISH

Robinson, Robert (1886–1975)

Perhaps the greatest of organic chemists in the classical tradition, famous for his research on natural products, especially alkaloids and plant pigments. Formulated an electronic theory of organic chemical reaction. Nobel Laureate 1947.

see CHESTERFIELD, LIVERPOOL, LONDON, MANCHESTER, OXFORD

Roe, Edwin Alliot Verdon (1877–1958)

The first Englishman to design, construct and fly his own aircraft. Founded A.V. Roe and Co. 1910, producing the AVRO 504 bomber/trainer.

see HAMBLE, LONDON, MANCHESTER

Roebuck, John (1718–94)

Pioneer industrial chemist: invented the lead-chamber process for sulphuric acid manufacture.

Worked initially in Birmingham but problems of maintaining secrecy led him to move to Prestonpans, near the Scottish bleachers.
see BO'NESS

Roget, Peter Mark (1779–1869)
Remembered mainly for his great *Thesaurus*, but also a scientist of note. First Fullerian Professor of Physiology, Royal Institution, London, 1833–6; invented a slide-rule: Secretary of the Royal Society 1827–49.
see GREAT MALVERN

Rolls, Charles (1877–1910)
Pioneer aviator and motorist: went into partnership with Henry Royce (q.v.) in 1906.
see MONMOUTH

Roscoe, Henry (1833–1915)
Professor of Chemistry, Owens College, Manchester 1855–87. First to isolate pure vanadium. A notable popularizer of science: wrote several very successful chemistry textbooks.
see MANCHESTER

Ross, Ronald (1857–1932)
Physician. Service in the Indian Medical Service led him to a keen interest in tropical diseases, especially malaria, and he worked out the life-cycle of the malaria parasite in birds. Nobel Laureate, with P. Manson (q.v.), 1902.
see LONDON

Royce, Henry (1863–1937)
Electrical and mechanical engineer. Founded Royce Ltd in Manchester in 1884, and went into partnership with C.S. Rolls (q.v.) in 1906.
see MANCHESTER

Russell, Edward John (1872–1965)
Agricultural scientist: Director of Rothamsted Experimental Station 1912–43. His interests were international and his *World Population and World Food Supplies* (1954) attracted wide attention.
see ABERYSTWYTH

Rutherford, Ernest (Lord Rutherford of Nelson 1871–1937)
Widely recognized as the founder of atomic physics. After research on magnetism in New Zealand a scholarship brought him to Cambridge where he investigated electrical conduction in gases. Appointed professor at McGill University, Montreal, in 1898 and did fundamental research on radioactivity. In 1907 moved to Manchester and began his revolutionary work on the structure of the atom. Evolved theory of

nuclear atom (1912) which was perfected by Niels Bohr (q.v.) invoking quantum theory. In 1919 appointed Cavendish Professor at Cambridge, inspiring a team of brilliant research workers. Remembered for the extreme simplicity of his apparatus.
see CAMBRIDGE, LONDON, MANCHESTER

Ryle, Martin (1918–84)
Astronomer and cosmologist. Wartime research on radar led him to an interest in the new science of radio astronomy. This he developed at Cambridge, cataloguing an ever-increasing number of radio stars. Nobel Laureate 1974.
see CAMBRIDGE, BRIGHTON

Sacrobosco (d.1244 or 1256)
One of the few certainties about the life of this great medieval scholar is that he is buried in the convent of St Mathurin, Paris. He left a legacy of works on mathematics and astronomy, some still used four centuries later.
see HALIFAX

Sadler, James (1753–1828)
Pioneer aeronaut and engineer.
see MANCHESTER, OXFORD

Sadler, John (1720–89)
Potter: introduced transfer-printing to the industry.
see LIVERPOOL

Salt, Titus (1803–76)
Wool manufacturer: first to make alpaca fabrics in England.
see BRADFORD

Sanger, Frederick (b.1918)
Biochemist, distinguished for his research on the structures of proteins and nucleic acid. The only person ever to receive two Nobel Prizes in chemistry (1958, 1980).
see RENDCOMBE

Savery, Thomas (1650–1715)
Inventor of steam pump for raising water 'by the Impellent Force of Fire'. It was not a steam engine in the true sense but his patents were so widely drawn that Thomas Newcomen (q.v.), inventor of the beam engine, had to defer to him.
see DARTMOUTH, LONDON

Savile, Henry (1549–1622)
Warden of Merton College, Oxford, 1585–1622. Founded prestigious chairs of geometry and astronomy.
see OXFORD

Schrödinger, Erwin (1887–1961)
Founder of wave mechanics. Author of the
Schrödinger Equation describing the behaviour of
electrons and other particles.
see DUBLIN

Scoresby, William (1789–1857)
Leader of several voyages of scientific exploration
to Australia and the Antarctic.
see WHITBY

Scott, Robert Falcon (1868–1912)
Antarctic explorer: died on return journey from the
South Pole 1912. His memorial is the Scott Polar
Research Institute, Cambridge.
see CARDIFF

Sedgwick, Adam (1785–1873)
Geologist, noted for his delineation of the Cambrian
system. The Sedgwick Museum of Geology,
Cambridge, is named after him.
see CAMBRIDGE, DENT

Shaw, William Napier (1854–1945)
Meteorologist: Director of the Meteorological
Office 1905–20.
see BIRMINGHAM

Sherrington, Charles (1857–1952)
Physiologist, renowned for his research on the structure and functioning of the nervous system. President
of the Royal Society 1920–5: Nobel Laureate 1932.
see CAMBRIDGE, OXFORD

Shrapnel, Henry (1761–1842)
Artillery officer: inventor of the type of explosive
shell that bears his name.
see BRADFORD-ON-AVON

Sibbald, Robert (1641–1722)
President of Edinburgh Royal College of Physicians
1684. A founder of the Botanical Garden, Edinburgh.
see EDINBURGH

Simpson, George (1878–1965)
Meteorologist. Revised Beaufort Scale of wind-speed
1926. Investigated generation of electricity in
thunderstorms. As meteorologist, sailed with R.F.
Scott (q.v.) to the Antarctic 1912.
see DERBY

Simpson, James (1811–70)
Physician: introduced chloroform as alternative
to ether in anaesthesia. Aroused much criticism
for his use of anaesthesia in childbirth, but

this faded after Queen Victoria accepted it from John
Snow (q.v.) at the birth of Prince Leopold (1853).
see EDINBURGH

Simson, Robert (1687–1768)
Professor of Mathematics, Glasgow, 1712–61.
see GLASGOW

Sloane, Hans (1660–1753)
Fashionable physician. His collection of over fifty
thousand books and manuscripts was the nucleus of
the British Museum.
see LONDON

Smeaton, John (1724–92)
Engineer and instrument maker. By systematic
application of scientific principles he effected
improvements in many machines, including wind
and water-mills. Also undertook many major civil
engineering projects.
see ABERDEEN, LEEDS, PLYMOUTH, ST AUSTELL

Smellie, William (1740–95)
Printer, antiquary and naturalist. Printed and
contributed to first edition of the *Encyclopaedia
Britannica* (1771).
see EDINBURGH

Smiles, Samuel (1812–1904)
Practised as surgeon in Leeds but forsook this for
journalism. Wrote many biographical works: his
Self-Help (1859) was a widely read guide to self-
improvement.
see LEEDS, BELFAST

Smiles, William (1846–1904)
Rope-maker, whose yard in Belfast became the
world's largest.
see BELFAST

Smith, William (1769–1839)
Geologist whose interest was aroused by the many
rock exposures he encountered in his countrywide
travels as a surveyor. His chief memorial is the large
geological map published in 1815 as *A Delineation
of the Strata of England and Wales.*
see CHURCHILL

Snow, John (1813–58)
Physician, remembered for having identified the
notorious Broad Street Pump in London as the
source of a cholera epidemic in 1854. His
administration of chloroform to Queen Victoria at
the birth of Prince Leopold (1853) stilled religious
criticism of the practice.
see LONDON

Soddy, Frederick (1877–1956)
Chemist, famous for formulating the concept of isotopes (1913). Nobel Laureate 1921.
see ABERDEEN, OXFORD

Somerville, Mary (1780–1872)
Mathematician and scientific writer. The breadth of her interest is revealed in *The Connection of the Physical Sciences* (1834).
see OXFORD

Sorby, Henry Clifton (1826–1908)
His interest in geology as a wealthy amateur led him to examine rocks in very thin sections under the microscope, earning him the title 'Father of Microscopical Petrography'.
see SHEFFIELD

Spencer, Herbert (1820–1903)
Practised as a civil engineer but abandoned this for writing. An advocate of 'social Darwinism', set out in his nine-volume *System of Synthetic Philosophy* (1862–93).
see DERBY

Sprat, Thomas (1635–1713)
One of the scientific circle which generated the Royal Society (1662) of which he was an original member. Wrote first history of the Society (1667).
see LONDON

Stephenson, George (1781–1848)
Railway engineer, pioneer of steam traction. Devised a miner's lamp which brought him into conflict with Humphry Davy (q.v.).
see KILLINGWORTH, LONDON

Stephenson, Marjory (1885–1948)
Biochemist, remembered for research, mainly on bacterial metabolism, under Sir Frederick Gowland Hopkins (q.v.) at Cambridge. With Kathleen Lonsdale (q.v.) one of first two women elected to Fellowship of the Royal Society.
see CAMBRIDGE, NEWBRIDGE

Stephenson, Robert (1803–1848)
Mechanical and civil engineer who collaborated with his father, George Stephenson (q.v.) in building loco-motives. Greatly involved also in the construction of railroads, including several major railway bridges.
see BANGOR, EDINBURGH, LONDON, NEWCASTLE UPON TYNE

Stirling, Robert (1790–1878)
Inventor of the Stirling heat engine, in which the expansive fluid is air rather than steam.
see GALSTON

Stoney, George (1826–1911)
Mathematical physicist, whose research interests ranged from optics to the atmospheres of the Sun and planets.
see DUBLIN, DUN LAOGHAIRE, GALWAY

Stringfellow, John (1799–1883)
Lace manufacturer who, with W.S. Henson (q.v.) had ambitious, but unrealized, plans for an 'aerial steam carriage'.
see CHARD

Strutt, Jedediah (1726–97)
Cotton-spinner: improved the stocking frame. Went into partnership with Richard Arkwright (q.v.).
see BELPER

Stubbs, George (1724–1806)
Painter, particularly remembered for his paintings of animals, especially horses. He studied anatomy at York and his classic *Anatomy of the Horse* appeared in 1766.
see LIVERPOOL

Sturgeon, William (1783–1850)
Private soldier who studied science, especially electricity, to such effect that he was appointed lecturer in science at the Royal Military College (1824).
see MANCHESTER

Sugden, Samuel (1892–1950)
Chemist: introduced the 'parachor', relating molecular volume to surface tension.
see LEEDS

Swan, Joseph (1828–1917)
Inventor, independently of Thomas Edison in the USA, of the incandescent filament lamp (1878). Keenly interested in photography, he introduced the then-revolutionary rapid dry plate.
see NEWCASTLE UPON TYNE, SUNDERLAND

Swinburne, James (1858–1958)
Engineer, much concerned with the generation and distribution of electricity. Pioneer of the plastics in-dustry, introducing phenol/formaldehyde resins in 1904.
see INVERNESS

Sydenham, Thomas (1624–84)
Celebrated physician. He had little faith in scientific aids to diagnosis, relying on meticulous clinical observation at the bedside – hence his sobriquet 'The

English Hippocrates'. First to describe Sydenham's Chorea (St Vitus' Dance).
see DORCHESTER

Synge, Richard Laurence Millington (1914–94)

Pioneer, with A.J.P. Martin (q.v.), of partition and paper chromatography. Nobel Laureate 1952.
see LEEDS

Szilard, Leo (1898–1964)

Atomic physicist; first to envisage the possibility of generating vast quantities of energy by a nuclear chain reaction, which he realized in Chicago, with Enrico Fermi, in 1942.
see LONDON

Talbot, William Henry Fox (1800–77)

Pioneer of photography, taking his first permanent photograph in 1835. His great contribution was the introduction of the negative, from which an unlimited number of positive prints could be made. Also a talented mathematician (FRS 1831) and deciphered many Assyrian cuneiform inscriptions.
see LACOCK

Telford, Thomas (1757–1834)

One of the great civil engineers of the Industrial Revolution, deeply involved in the building of roads, canals and bridges. His greatest achievement was the improved London–Holyhead road.
see ABERDEEN, LLANGOLLEN, LONDON, LONGTON UPON TERN

Tennant, Charles (1768–1838)

Chemical industrialist, who revolutionized the textile industry by introducing bleaching powder. In 1835 his 100-acre factory at St Rollox was the largest chemical plant in the world.
see GLASGOW

Thomas, Sydney Gilchrist (1850–85)

Metallurgist who, with his cousin Percy Carlyle Gilchrist (q.v.), devised a process for smelting the world's vast deposits of phosphoric iron ore, unworkable by processes then available.
see BLAENAVON

Thompson, Benjamin (Count Rumford 1753–1814)

American-born social scientist, founder of the Royal Institution in London, who established a quantitative relationship between work and heat. Married Lavoisier's widow. Highly regarded in America:

Thomas Jefferson regarded him as 'the greatest mind that America has produced'.
see LONDON

Thompson, D'Arcy Wentworth (1860–1948)

Zoologist. Published his influential *On Growth and Form* in 1917.
see EDINBURGH

Thomson, Charles Wyville (1830–82)

Marine biologist: led the great circumnavigational voyage of the *Challenger* 1872–6.
see EDINBURGH

Thomson, Joseph John (1856–1940)

One of the founders of the British school of atomic physics. Investigated discharge of electricity through gases, leading to discovery of electrical 'corpuscles' in 1899, subsequently named electrons by G.J. Stoney (q.v.).
see CAMBRIDGE, LONDON

Thomson, William (Baron Kelvin of Largs 1824–1906)

One of the great physicists of the 19th century, making major advances in electricity, magnetism and heat. Made important improvements in a number of instruments, including the marine compass.
see GLASGOW

Thorburn, Archibald (1860–1935)

Artist famous for his meticulously accurate paintings of birds and animals.
see DOBWALLS

Tinbergen, Nikolas (1907–88)

Dutch-born zoologist, co-founder of the science of ethology, the objective study of animal behaviour. Nobel Laureate 1973.
see OXFORD

Tite, William (1798–1873)

Architect and civil engineer, whose projects included London main-line stations and the Thames Embankment, the Royal Exchange and docks.
see SOUTHAMPTON

Todd, Alexander (Lord Todd of Trumpington, b.1907)

Organic chemist, with notable achievements in the chemistry of natural products, including vitamins, antibiotics, blood anticoagulants and nucleic acids. Nobel Laureate 1957.
see CAMBRIDGE, GLASGOW, MANCHESTER

Tompion, Thomas (1639–1713)

Famous English horologist: clockmaker to the newly founded Royal Observatory 1676. Master of the Clockmakers Company 1703.

see BEDFORD, NORTHILL

Townshend, Charles (Viscount Townshend 1674–1738)

Agriculturalist. Improved productivity of land by introducing the turnip into the regular agricultural programme, hence the sobriquet 'Turnip' Townshend.

see RAYNHAM

Tradescant, John (1570–1638)

Gardener, who established a physic garden in Lambeth. On his death his collection became the nucleus of the Ashmolean Museum, Oxford. Genus Tradescantia named for him.

see LONDON

Travers, Morris William (1872–1961)

Chemist: collaborated with William Ramsay (q.v.) in the isolation of the rare gases of the atmosphere.

see BRISTOL

Trevithick, Richard (1771–1833)

Pioneer of high-pressure steam. Invented a successful double-action steam engine for pumping water from mines and other purposes. Also built road and rail locomotives. Died penniless.

see CAMBORNE, LONDON, MERTHYR TYDFIL

Turner, William (c. 1508–68)

Botanist whose religious beliefs obliged him to spend much of his life abroad. Best-known for his three-volume Herball (1551–68), which included much original information. Also wrote a book on birds: Avium Praecipuarum (1554).

see MORPETH

Turner, William (1881–1963)

Pioneer of the scientific study of glass.

see SHEFFIELD

Twort, Frederick (1877–1950)

Made important contributions to microbiology, notably his discovery (1915) of bacteriophage, a virus which consumes bacteria.

see CAMBRIDGE

Tyndall, John (1820–93)

In 1843 succeeded Michael Faraday (q.v.) at the Royal Institution. Investigated properties of crystals in a magnetic field and action of radiant heat on gases. Remembered for Tyndall Effect, resulting from the differential scattering of light by fine particles in the air.

see LEIGHLINBRIDGE

Ussher, James (1581–1656)

Archbishop of Armagh. His place in cosmology assured by his confident estimate, on Biblical rather than scientific evidence, that the world was created in 4004 BC.

see ARMAGH

Wakeley, Thomas (1795–1862)

Medical reformer: founded the Lancet 1823.

see LONDON

Walker, John (1780–1859)

Practised as a surgeon, but set up in business as an apothecary. Inventor of the first friction match, about 1827.

see STOCKTON-ON-TEES

Wallace, Alfred Russel (1823–1913)

Naturalist. Conceived, independently of Charles Darwin (q.v.), a theory of organic evolution by natural selection. They made a joint presentation to the Linnaean Society, London, 1 July 1858.

see BROADSTONE, LONDON, USK

Wallingford, Richard of (c. 1291–1336)

The leading mathematician and astronomer of his day: builder of one of the first mechanical clocks.

see WALLINGFORD

Wallis, John (1616–1703)

Savilian Professor of Geometry, Oxford, 1649–1703. Wrote widely on mathematics, including cryptography. Published a well-known formula for calculating π.

see ASHFORD, CHESTER, LONDON, OXFORD

Walton, Ernest (b.1903)

Conducted with J.D. Cockcroft (q.v.) the famous atom-splitting experiment in Cambridge 1932. Nobel Laureate 1951.

see DUBLIN

Watson, James (b.1928)

Molecular biologist: he was joint discoverer, with Francis Crick (q.v.), of the structure of DNA – the Golden Helix. Nobel Laureate 1962.

see CAMBRIDGE

Watson-Watt, Robert (1892–1973)

Pioneer of radar development, beginning his

research in 1935. By the outbreak of war in 1939 Britain had a chain of radar stations, which was of critical importance in the Battle of Britain in 1940.
see BRECHIN, DUNDEE, ORFORD

Watt, James (1736–1819)

One of the greatest engineers of the Industrial Revolution, inventor of the improved steam engine (1764). Partner with Matthew Boulton (q.v.) at the famous Soho factory. A prominent member of the Lunar Society.
see BIRMINGHAM, GLASGOW, GREENOCK

Wedgwood, Josiah (1730–95)

A great potter, who experimented tirelessly to produce improved wares. In this he was much helped by the advice of scientific friends in the Lunar Society. It was said that 'he elevated pottery manufacture . . . to a matter of scientific measurement and calculation'.
see BIRMINGHAM, STOKE-ON-TRENT

Weizmann, Chaim (1874–1952)

Organic chemist: ardent Zionist who became first President of Israel.
see LONDON, MANCHESTER

Weldon, Walter (1832–85)

Began life as a journalist, founding *Weldon's Journal*, a popular fashion magazine. Taught himself chemistry and devised a novel process for improving soda manufacture. According to Jean Dumas: 'Every sheet of paper and every yard of calico has been cheapened throughout the world.'
see LOUGHBOROUGH

Wellcome, Henry (1853–1936)

Highly successful drug manufacturer: philanthropist. Founded the Wellcome Trust, now the largest private supporter of medical research in Britain. His enormous collection of medical artefacts is now in the Science Museum.
see LONDON

Wheatstone, Charles (1802–75)

Physicist: remembered for commercial development of the electric telegraph with W.F. Cooke (q.v.).
see GLOUCESTER, LONDON, SWANSEA

Wheeler, Robert Eric Mortimer (1890–1976)

Distinguished archaeologist: Director of the London Museum 1926–44. Remembered for his excavation

of Roman sites in Britain and at Mohenjo-daro and Harappa in India.
see STANWICK

Whewell, William (1794–1866)

Successively Professor of Mineralogy and Moral Theology at Cambridge. Wrote variously on electricity and magnetism, and on the tides. Coined the word 'scientist' in its modern sense.
see CAMBRIDGE, LANCASTER

Whipple, Robert (1871–1953)

Chairman, Cambridge Instrument Co. 1939–49. His collection of historic instruments and books is the basis of the Whipple Museum, Cambridge.
see CAMBRIDGE

Whitehead, Robert (1823–1905)

Engineer. Manufactured land-drainage and textile machinery, but remembered chiefly for invention of the torpedo (1866).
see BOLTON

Whitehurst, John (1713–88)

Manufacturer of clocks and scientific instruments.
see DERBY

Whittle, Frank (b.1907)

Engineer: British pioneer in the development of jet engines for aircraft.
see CAMBRIDGE, COVENTRY

Whitworth, Joseph (1803–87)

Engineer, remembered for his introduction of new standards of precision into the machine tool industry. In 1830 accuracy of 1/16th inch was reckoned good: Whitworth set 1/10,000th inch as his workshop standard, though he could measure even more accurately.
see MANCHESTER

Wilberforce, Samuel (1805–73)

'Soapy Sam': Bishop of Oxford 1845–69. Remembered for famous clash with T.H. Huxley (q.v.), defending Darwin's theory of evolution.
see OXFORD

Wilkins, John (1614–72)

Mathematician: a founder member of the Royal Society. Versatile writer: wrote on codes and ciphers and made an imaginative study of life on the Moon.
see OXFORD

Wilkinson, John (1728–1808)

Engineer, whose most important invention (1774)

was his boring machine, used for the construction of cylinders needed for Boulton and Watt steam engines.
see BERSHAM, CLIFTON, PENRITH

Willett, William (1856–1915)
Builder, powerful advocate of Daylight Saving, introduced 1916.
see PETT'S WOOD

William of Sherwood (b.1205)
Logician: his life, like that of many peripatetic medieval scholars, is obscure and he is remembered by his writings. Roger Bacon (q.v.) called him 'one of the famous wise men in Christendom'.
see NOTTINGHAM

Willis, Thomas (1621–75)
Physician, famous for his account of the anatomy of the brain. First to diagnose diabetes mellitus by sweetness of the urine.
see GREAT BEDWYN

Williams, Thomas (1737–1802)
Welsh mining entrepreneur, who developed the vast copper ore deposits at Parys Mountain, Anglesey.
see AMLWCH

Willughby, Francis (1635–72)
Naturalist: collaborated with John Ray (q.v.) until his untimely death.
see BRAINTREE

Wilson, Charles Thomson Rees (1869–1959)
Atomic physicist: remembered for his invention of the Wilson Cloud Chamber for detecting the tracks of atomic particles.
see GLENCORSE

Winsor, Frederick (1763–1830)
German entrepreneur: a pioneer of gas-lighting in Britain.
see LONDON

Withering, William (1741–99)
A fashionable physician in Birmingham and a member of the influential Lunar Society. As a botanist remembered for his systematic survey of plants, *The Botanical Arrangement* (1776). Introduced the use of foxglove to treat dropsy.
see BIRMINGHAM

Wolff, Gustav (1834–1913)
A founder of the great Harland and Wolff shipbuilding company.
see BELFAST

Wollaston, William (1766–1828)
Chemist, the first to produce platinum in a malleable and useful form. Discovered rhodium and palladium.
see CAMBRIDGE, EAST DEREHAM, LONDON

Wood, John (1705–1754)
Architect, noted for his designs in the Palladian style.
see BATH

Woulfe, Peter (1727–1803)
Chemist and mineralogist: discovered native tin in Cornwall 1766.
see DUBLIN

Wren, Christopher (1632–1723)
Architect and mathematician: a member of the scientific circle from which the Royal Society evolved in 1662: President 1680–2. Savilian Professor of Astronomy, Oxford, 1661–73.
see LONDON, OXFORD, WARMINSTER

Wright, Joseph (1734–97)
Painter, whose works often depict scientific themes.
see DERBY

Wright, Thomas (1711–86)
Astronomer, who had a varied career as teacher of navigation, lecturer in science and land surveyor to the gentry. Sought in his writings to reconcile science and religion.
see BYERS GREEN

Young, James (1811–83)
Industrialist: founded the Scottish shale-oil industry.
see GLASGOW

Young, Thomas (1773–1810)
An infant prodigy, gifted with phenomenal memory, who established the wave, as opposed to the corpuscular, theory of light. Latterly devoted much time to deciphering Egyptian hieroglyphics, including those of the Rosetta Stone.
see MILVERTON

FURTHER READING

A large number of reference books – including standard histories of science and technology – have been drawn upon in compiling this gazetteer. Many are specialized but the following, of a more general nature, are suggested for complementary reading.

Berry, Elizabeth. *The Writing on the Wall: A Guide to the Inscriptions on Edinburgh Buildings Which Commemorate Famous Citizens, Visitors and Events*, Edinburgh: Cockburn Association, 1990

Cossons, Neil. *The BP Book of Industrial Archaeology in Britain* (2nd ed.), David and Charles, 1987

Dictionary of National Biography, Oxford University Press, 66 vols and supps, 1885–

Dictionary of Scientific Biography, Scribners, 18 vols, 1970–90

English Heritage. *The Blue Plaque Guide (London)*, Journeyman Press, 1991

Greenwood, Douglas. *Who's Buried Where in England*, Constable, 1982

Hall, A. Rupert. *The Abbey Scientists: The Memorials of Westminster Abbey*, R. & N. Nicholson, 1966

Hayes, Andrew. *Archaeology of the British Isles*, Batsford, 1993

Hibbert, Christopher and Hibbert, Edward (eds). *The Encyclopaedia of Oxford*, Macmillan, 1988

Holbrook, Mary. *A Directory of Scientific Instruments in Collections in the United Kingdom and Eire*, HMSO, 1992

Hudson, Kenneth and Nicholls, Ann. *Cambridge Guide to the Museums of Great Britain and Ireland*, Cambridge University Press, 1989

Lines, Clifford. *Companion to the Industrial Revolution*, Facts on File, 1990

Liverpool Heritage Walk, Liverpool City Council, 1990

Men and Women of Manchester, City of Manchester Public Relations Office, 1979

Map of Ancient Britain: A Map of the Major Visible Antiquities of Great Britain Older than A.D. 1066, Ordnance Survey, 1964

Minchinton, W. *A Guide to Industrial Archaeology Sites in Great Britain*, Paladin, Granada, 1984

Museums and Galleries in Great Britain and Ireland, British Leisure Publications, 1990

Musson, A.E. and Robinson, Eric. *Science and Technology in the Industrial Revolution*, Manchester University Press, 1969

Rosen, Dennis and Sylvia. *London Science: Museums, Libraries and Places of Scientific, Technological and Medical Interest*, Prion, 1994

Royal Commission on the Historical Monuments of England. *Industry and the Camera*, HMSO, 1985

Shipley, Debra and Peplow, Mary. *The Other Museum Guide*, Grafton, 1988

Tanford, Charles and Reynolds, Jacqueline. *A Travel Guide to Scientific Sites of the British Isles*, Wiley, 1995

Trinder, Barrie, *Industrial Heritage of Britain*, Automobile Association, 1992

Tyson, Colin. *The Great British Steam Railway Timetable*, Alan Sutton, 1996

Weinreb, Ben and Hibbert, Christopher (eds). *The London Encyclopaedia*, Macmillan, 1983

Williams, Trevor I. (ed.). *A Biographical Dictionary of Scientists* (4th ed.), HarperCollins, 1994